Rare-Earth Metal Hexaborides: Synthesis, Properties, and Applications

Authored by

Mikail Aslan

Gaziantep University, Faculty of Engineering
Department of Metallurgical and Material Science
Engineering
Gaziantep, 27310
Turkey

&

Cengiz Bozada
Gaziantep University, Faculty of Engineering
Department of Physics
Gaziantep, 27310
Turkey

Rare-Earth Metal Hexaborides: Synthesis, Properties, and Applications

Authors: Mikail Aslan and Cengiz Bozada

ISBN (Online): 978-981-5124-57-6

ISBN (Print): 978-981-5124-58-3

ISBN (Paperback): 978-981-5124-59-0

© 2023, Bentham Books imprint.

Published by Bentham Science Publishers Pte. Ltd. Singapore. All Rights Reserved.

First published in 2023.

need for a court order if at any point you breach any terms of this License Agreement. In no event will any delay or failure by Bentham Science Publishers in enforcing your compliance with this License Agreement constitute a waiver of any of its rights.

3. You acknowledge that you have read this License Agreement, and agree to be bound by its terms and conditions. To the extent that any other terms and conditions presented on any website of Bentham Science Publishers conflict with, or are inconsistent with, the terms and conditions set out in this License Agreement, you acknowledge that the terms and conditions set out in this License Agreement shall prevail.

Bentham Science Publishers Pte. Ltd.
80 Robinson Road #02-00
Singapore 068898
Singapore
Email: subscriptions@benthamscience.net

BENTHAM SCIENCE

CONTENTS

PREFACE

Rare-earth hexaborides have attracted continuous attention for more than half a century, both from the point of view of fundamental material sciences and for practical applications in various fields of engineering. These materials indicate a wealth of unusual electronic, mechanical, optical, and magnetic properties that have been closely investigated in recent decades using advanced spectroscopies and state-of-the-art physical characterization methods.

This book consists of a comprehensive collection of reviews offering a cutting-edge summary of the investigations based on rare-earth hexaborides from various viewpoints. The book includes chapters on the growth and characterization of different structure types of rare-earth hexaborides, and their theoretical and experimental descriptions, production methods, unusual properties, and improvements by alloying and compositing.

The book will appeal to anyone interested in material science, physics, and chemistry, especially researchers and postgraduate students who focus on production methods, structure types, and applications of rare-earth hexaboride compounds.

CONSENT FOR PUBLICATION

Not applicable.

CONFLICT OF INTEREST

The authors declare that they have no conflict of interest.

Mikail Aslan
Gaziantep University, Faculty of Engineering
Department of Metallurgical and Material Science Engineering
Gaziantep, 27310
Turkey

&

Cengiz Bozada
Gaziantep University, Faculty of Engineering
Department of Physics
Gaziantep, 27310
Turkey

ACKNOWLEDGEMENTS

I would like to express my sincere gratitude and appreciation to many people who have made this book possible. Amongst them, I would like to express the most profound and sincere gratitude to my family for his priceless support, encouragement, understanding, patience, great help, and motivation throughout this bookwork and my academic life. I would like to thank all members of Gaziantep University for guiding and encouraging me during my academic life.

This book was supported by "Scientific Research Governing Unit of Gaziantep University in Turkey" with the project title of "The production of high purity metal hexaborides consisting of rare-earth elements used in advanced technological applications and improvement of their mechanical, thermal and optical properties" and the project number of MF.DT.20.06.

CHAPTER 1

The Rare-Earth Elements

Abstract: In this section, the elemental forms of rare-earth elements are iron gray to silvery lustrous metals that are typically soft, malleable, ductile, and usually reactive, especially at elevated temperatures or when finely divided. rare-earth elements are examined in terms of physical and chemical properties. This makes them essential components of diverse defense, energy, industrial, military technology, and low-carbon technologies. Furthermore, REEs are rapidly being used in magnet applications. For example, magnets produced by Neodymium-iron, the strongest known type of magnet, are used widely. Thus, their application areas vary from the electronic to glass industry. Also, information about the sources of rare-earth elements is given in this part.

Keywords: Light rare-earth elements, Heavy rare-earth elements.

1.1. INTRODUCTION

Rare-earth elements (REEs) consist of a group of 15 elements between Lanthanum and Lutetium. Based on their atomic mass, they are generally classified as light and heavy REEs (light rare-earth elements: Lanthanum (La), Cerium (Ce), Praseodymium (Pr), Neodymium (Nd), promethium (Pr), and Samarium (Sm), and heavy rare-earth elements: Europium (Eu), Gadolinium (Gd), Terbium (Tb), Dysprosium (Dy), Holmium (Ho), Erbium (Er), Thulium (Tm), Ytterbium (Yb) and Lutetium (Lu)). REEs are a group of chemically similar elements with atomic numbers from 57 to 71. Yttrium and Scandanium are 39 and 21 atomic numbers, respectively. They have also been recently regarded as REEs since they share chemical and physical similarities and have affinities with the Lanthanides [1 - 13]. The members of REEs are given in Fig. (**1.1**).

The principal economic sources of REEs are the minerals: bastnasite, monazite, loparite, and the lateritic ion-adsorption clays. rare-earth is a relatively abundant group of 17 elements composed of scandium, yttrium, and lanthanides. The elements range in crustal abundance from cerium, the most abundant element of the 78 common elements in the Earth's crust at 60 parts per million, to thulium and lutetium, the least abundant rare-earth elements at about 0.5 parts per million.

Mikail Aslan and Cengiz Bozada

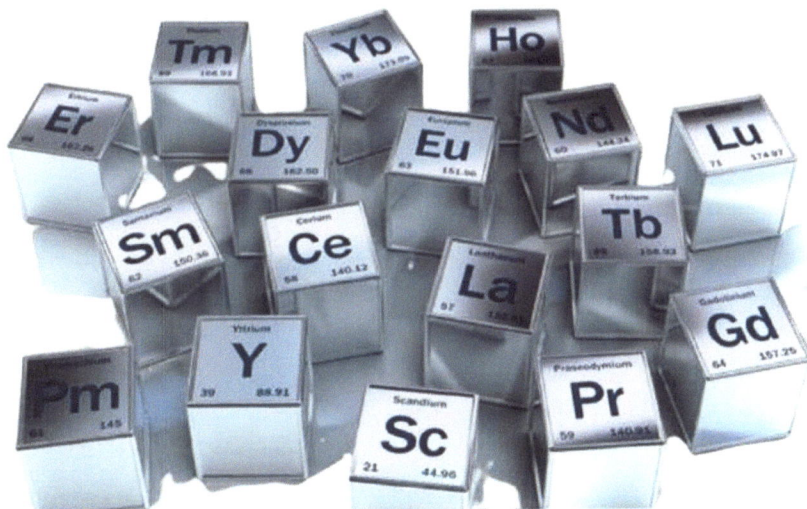

Fig. (1.1). The lists of rare-earth elements.

The elemental forms of REEs are iron gray to silvery lustrous metals that are typically soft, malleable, ductile, and usually reactive, especially at elevated temperatures or when finely divided.

The REEs have unusual physical and chemical properties, making them essential components of diverse defense, energy, industrial, military, and low-carbon technologies. The REE raw materials are widely consumed in the glass industry for glass polishing and as additives providing color and special optical properties to the glass. Lanthanum and cerium-based catalysts are preferred in petroleum refining and automotive catalytic converters, respectively. REEs are rapidly being used in magnet applications. For example, magnets produced by Neodymium-iron, the strongest known type of magnet, are used widely. Nickel-metal hydride batteries use anodes made of lanthanum-based alloys.

In this part, we have focused on the properties and the application areas of REEs, which will be discussed in detail in the following subchapters.

1.2. LIGHT RARE-EARTH ELEMENTS

1.2.1. Scandium (Sc)

Scandium (Sc) is in the IIIB group and is the lightest element of transition metals. Scandium is a white-silver metal. The atomic number of scandium is 21, and its atomic weight is 44.95 g/mol. It is a very hard-to-obtain, expensive, but precious element. Scandium which is between (REEs) and transition metals, increases the hardness of the material considerably, although it is added to the materials at a very small ratio. The properties of Sc are summarized in Table **1.1**.

Table 1.1. The properties of Scandium (Sc) [14].

Atomic weight	44.9559 g/mol
Pauling electronegativity scale	unknown
Intensity	3.0 g/cm^3 at 20 °C
Liquefaction point	1541 °C
Simmer point	2836 °C
Intermolecular forces	0.161 nm
Ionic radii	0.083 nm (+3)
Nuclide	7
Main energy level	[Ar] 3d^1 4s^2
First ionization energy	640.5 kJ/mol
Second ionization energy	1233 kJ/mol

It is used as a hardness-enhancing material in the body parts of bicycles, baseballs, and golf vehicles (Fig. **1.2**). It is also used in aviation, which includes warplanes. Recently, this element has been used as an important light source in high-quality lamps [13]. Generally,

• Scandium element is used in the production of powerful light bulbs used in night lighting and also has a daylight effect,

• Scandium-aluminium alloys are used in aircraft body production in terms of the lightness of warplanes and better maneuverability (Fig. **1.2**).

• Scandium-aluminum is used for the production of bicycle bodies due to its strong and lightweight,

• Gadolinium-scandium-gallium-garnet crystals are used in the production of defense materials and devices,

• Yttrium-scandium-gallium garnet laser is used for root canal treatments in dentistry,

• Scandium-aluminium is used in weapons production because it is light and resistant [15].

Fig. (1.2). Uses of Scandium **a)** Warplane **b)** Bicycle **c)** Jet engine.

1.2.2. Yttrium (Y)

Yttrium (Y), with proton number 39 and an atomic mass of 88.92 g/mol, is a glossy silvery metal. The properties of Y are summarized in Table **1.2**. It is relatively stable in the air. Y is employed as a catalyst for definite reactions. Y is formed in uranium ores, but not found in pure form. That element is hard to divide from other REEs. Commercially, Y is fabricated by decreasing fluoride with calcium metal, but it could be fabricated through other processes. In nature, it is seen in the form of a dark gray powder. It can blaze in the air at temperatures overheating 400 °C. It is used in various alloys. Electron receiver properties are used to ensure the emptying of electron tubes. Color televisions are displayed through a display coated with yttrium oxide.

Table 1.2. The properties of Yttrium (Y).

Proton number	39
Atomic weight	88.9059 g/mol
Pauling electronegativity scale	1.2
Intensity	4.47 g/cm^3 at 20 °C
Liquefaction point	1500 °C
Simmer point	3336 °C
Intermolecular forces	0.106 nm (+3)
Nuclide	10
Main energy level	[Kr] 4d^1 5s^2
First ionization energy	626 kJ/mol
Second ionization energy	1185 kJ/mol
Third ionization energy	1980 kJ/mol

Generally;

• Used in yitrium-iron crystals in radars and microwave-operated devices.

• Used for disintegrating grain size in metals, such as titanium, molybdenum, chromium, and zirconium,

• Used to strengthen aluminum and magnesium mixtures,

• Used to prevent oxidation in vanadium and similar metals,

• Used in the jewelry industry (Fig. **1.3**),

Fig. (1.3). Uses of Yttrium (Y) **a)** Jewelry **b)** Ceramics **c)** Decoration.

• Used in glass and ceramic production,

• Color televisions are displayed through a display coated with yttrium oxide,

• Used in fire-resistant brick,

• Used in laser systems and camera lenses [16].

1.2.3. Lanthanum (La)

Lanthanum is in the IIIB group on the periodic table and is a silvery-white metallic element. The atomic number of the lanthanum element is 57, and its atomic weight is 138.92 g/mol. La is such a malleable material that quickly oxidizes if exposed to air. La has a hexagonal crystal structure which is a forgeable, silvery-white metal. Lanthanum is most usually produced from bastnäsite and monazite. It has the characteristics of burning with a very bright flame, being affected by dilute acids, darkening in non-humid environments, and reacting very quickly with hot water. The properties of La are summarized in Table **1.3**. They react directly with phosphorus, selenium, sulfur, carbon, boron, nitrogen, and halogens. The element La is trivalent, and its salts are colorless. They are found in solid form in nature [17]. Generally,

Table 1.3. The properties of Lanthanum (La).

Proton number	57
Atomic weight	138.91 g/mol
Pauling electronegativity scale	1.1
Intensity	6.18 g/cm^3 at 20 °C
Liquefaction point	920 °C
Simmer point	3464 °C
Intermolecular forces	0.161 nm
Ionic radii	0.104 nm (+3)
Nuclide	7
Main energy level	(Xe) 5d^1 6s^2
First ionization energy	539 kJ/mol
Second ionization energy	1098 kJ/mol
Third ionization energy	1840 kJ/mol
Standard potential	-2.52 V

• Used in the carbon-based lighting industry and studio lighting and projection machines (Fig. **1.4**),

• It is used to increase the data transfer speed of fiber optic cables, high-resolution cameras, telescopes, night vision binoculars, and qualified camera lenses,

• La_2O_3 is used in the construction of special optical glass as it increases the alkaline resistance of glass,

• Used as a small number of additive materials in the production of granular cast iron,

• Used as an additive material in fuel,

• Used in hydrogen-absorbent sponge alloys. These alloys can provide heat energy for each hydrogen absorption,

• Used for pH adjustment in swimming pools,

• Used in oil refineries [15].

Fig. (1.4). Uses of Lanthanum **a)** Projection machines **b)** Telescope **c)** Camera lens.

1.2.4. Cerium (Ce)

Cerium (Ce) is the most abundant element in REEs. It is the 25th most abundant in the earth's crust. When exposed to air, it darkens, burns even scratched with a knife, reacts quickly with water, and dissolves in acids. Ce is affected by air, becomes dull, and forms a layer of oxide that wears out on its surface, like the corrosion of iron. The properties of Ce are summarized in Table **1.4**. The atomic number of Cerium is 58 and its atomic weight is 140.1g/mol. Ce has a silver-bright color. In nature, there are found in very small ratios in some mines' compounds. Pure Ce is as soft as tin and easily processed. Although pure Ce is grey glossy, it quickly gets a dull color in the air. It's a powerful reduction in a chemical reaction. It reacts with water slowly in the cold and quickly in the heat. When Ce interacts with flame, it is oxidized, so the insulation of the element cerium is very difficult. Ce is moderately toxic. It spontaneously ignites at 65-80 °C. The flaming Ce shouldn't be extinguished with water because it reacts with water and creates hydrogen gases. One of the warnings is its reaction with zinc since it causes explosions [17, 18].

Table 1.4. The properties of Cerium (Ce).

Proton number	58
Atomic weight	140.12 g/mol
Pauling electronegativity scale	1.1

(Table 1.4) cont.....

Intensity	6.76 g/cm^3 at 20 °C
Liquefaction point	799 °C
Simmer point	3426 °C
Intermolecular forces	0.181 nm
Ionic radii	0.102 nm (+3)
Nuclide	9
Main energy level	(Xe) 4f^1 5d^1 6s^2
First ionization energy	526.8 kJ/mol
Second ionization energy	1045 kJ/mol
Third ionization energy	1945.6 kJ/mol
Fourth ionization energy	3537 kJ/mol
Standard potential	-2.48 V

Generally;

•It is used as metal in lighting in cinema, television, and similar industrial products (Fig. **1.5**),

• It is also used in the production of fluorescent bulbs,

• It is widely used to increase the oxidation resistance of superalloys in high temperatures,

• It is also used as a color remover during the production of thinning of glass,

• Cerium salts are performed in the photography and weaving industry,

• It is used as porcelain coatings are used for non-transparency,

• It is used in the carbon electrodes of arc lamps,

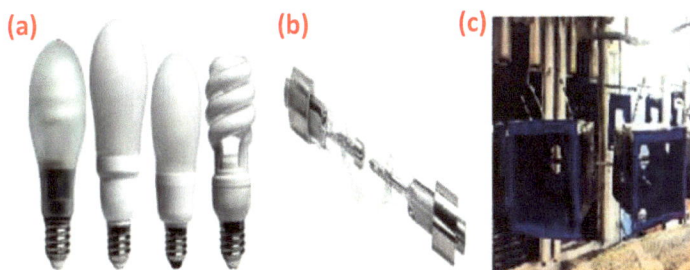

Fig. (1.5). Uses of Cerium **a)** Fluorescent bulb **b)** Arc lamp **c)** Porcelain coatings.

1.2.5. Praseodymium (Pr)

Praseodymium (Pr) is a soft, and glittery yellow metal. Praseodimun is a pale yellow element with proton number 59 and an atomic mass of 140,92 g/mol. It is in the IIIB group of the 6th period on the periodic table. The properties of Sc are summarized in Table **1.5**. Enamel and glass materials are given yellow color by Pr. Pr is quite forgeable and flexible. Pr is an interesting element in terms of paramagnetic at entire temperatures over 1K. Other REEs are antiferromagnetic or ferromagnetic at low temperatures. As a result of alloying Pr with another metal, high-strength metals, strong rare-earth magnets, and magnetocaloric materials are produced. At room temperature, they react gradually with oxygen, and if subjugated to air, they form a green oxide. They are more resistant to corrosion compared to other REEs. They react quickly with water [18]. Generally,

Table 1.5. The properties of Praseodymium (Pr).

Proton number	59
Atomic weight	140.91 g/mol
Pauling electronegativity scale	1.1
Intensity	6.8 g/cm^3 at 20 °C
Liquefaction point	931 °C
Simmer point	3512 °C
Intermolecular forces	unknown
Ionic radii	0.101 nm (+3)
Nuclide	5
Main energy level	(Xe) 4f^5 6s^2
First ionization energy	522 kJ/mol
Second ionization energy	1016 kJ/mol
Third ionization energy	2082.4 kJ/mol
Fourth ionization energy	3752 kJ/mol
Fifth ionization energy	5534 kJ/mol
Standard potential	-2.47 V

• Used to make high-power magnets with neodymium (Fig. **1.6**),

• Used in the production of welder and glass blowing goggles with yellow flare and protective glass against ultraviolet rays,

• Used in lighting and projection in film workshops,

• Praseodimium salt is used in enamel and glass colorings, making live yellow porcelain dining sets,

• Used in protective goggles and glass making used by welders,

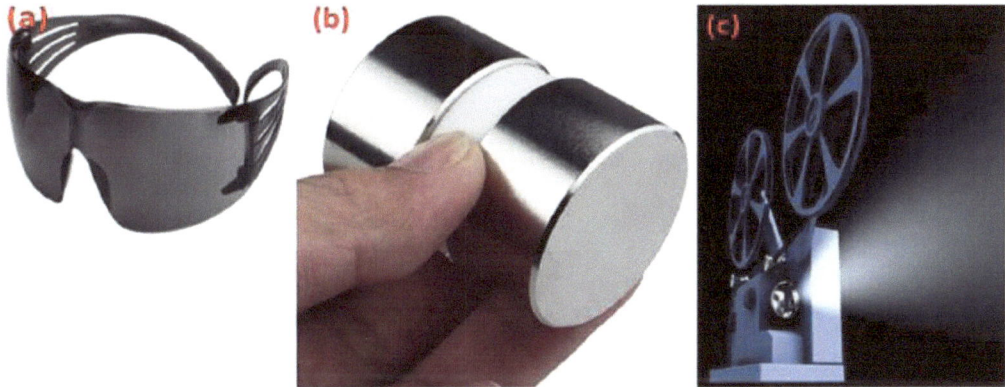

Fig. (1.6). Uses of Praseodymium **a)** Protective glasses **b)** Strong magnet **c)** Projection.

The ability to filter the yellow light that occurs during the glass manufacturing process, therefore, protection is used in the production of glasses for glass workers [15].

1.2.6. Neodymium (Nd)

Neodymium (Nd) is a soft, shiny, silvery-white metal with a proton number of 60 and an atomic mass of 144,3 g/mol. Nd is an element under the classification of lanthanides in the periodic table. It is the second element in terms of abundance among lanthanites found in nature. It forms an oxide coating in the air. The properties of Nd are summarized in Table **1.6**. Nd reserves are around 8 million tons, and the production of neodymium oxide is about 7,000 tons per year. Nd in humans is small, and although metal has no biological role, parts of the body may be affected. Nd dust and salts irritate the eyes a lot. Ingested Nd salts are a bit toxic if they dissolve, but not toxic if they are not dissolved. Nd is usually dangerous in the working environment because moisture and gases can be inhaled by air. When exposed to Nd for a long time, it can cause lung embolisms. If Nd material accumulates in the human body, it can pose a risk to the liver.

Table 1.6. The properties of Neodymium (Nd).

Proton number	60
Atomic weight	144. 2 g/mol

Pauling electronegativity scale	1.14
Intensity	7.0 g/cm^3 at 20 °C
Liquefaction point	1024 °C
Simmer point	3074 °C
Intermolecular forces	0.181 nm
Ionic radii	9
Nuclide	(Xe) 4f^1 6s^2
Main energy level	533 kJ/mol
First ionization energy	1040 kJ/mol
Second ionization energy	2130 kJ/mol

Generally;

• Neodymium is used to make strong magnets. For example, Neodymium- Iron-Boron ($Nd_2Fe_{14}B$) is one of the strongest magnets (Fig. **1.7**).

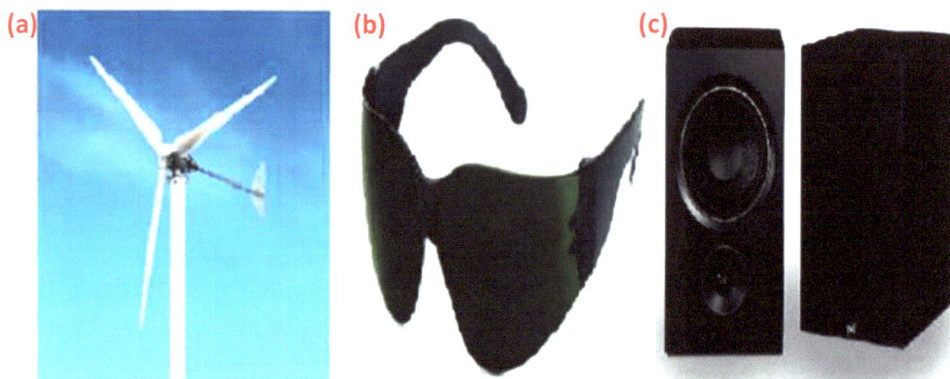

Fig. (1.7). Uses of Neodymium (Nd) **a)** Wind power generator **b)** Glasses for filtering infrared rays **c)** Loudspeaker.

• These magnets are used in wind power generators, electric vehicle engines, headphones, and speakers.

• Used as crystalline (neodymium-additive yttrium aluminum garnet) in lasers.

• It is a color ingredient that causes shades in a variety of colors, from purple to red and gray, on the glass surface. The sparkle passing through the surface of the glass shows very serious absorption tapes. This can also be used in astronomy fields.

• Neodymium is used in the production of glass that filters bright purple-colored glass and infrared rays.

• Together with Praseodim, they are used in the production of protective goggles used by glass manufacturers and welders.

• Neodymium salts are used in enamel colorings [19].

1.2.7. Promethium (Pm)

Promethium (Pm) is an element in the group of REEs with atomic number 61 and an atomic weight of 145 g/mol. Pm is not found in observable amounts in the world. The properties of Pm are summarized in Table **1.7**. It can be found in small amounts of uranium or as a thruput of uranium deterioration. It can also be produced as a product of uranium fission. That's why it's an element that's artificially produced in nuclear reactors. It is a colorless element found in solid form in room conditions. It has malleable and soft mechanical properties. Pm shows radioactive properties in REEs. In the dark, its salts glow with a light blue or green glow because of their high radioactivity. Pm could also be used as a beta source for some instruments used for thickness measurements. Pm is also used as an energy source for atomic batteries in spacecraft and guided missiles.

Table 1.7. The properties of Promethium (Pm).

Proton number	61
Atomic weight	147 g/mol
Pauling electronegativity scale	Unknown
Intensity	6.475 g/cm^3 at 20 °C
Liquefaction point	1168 °C
Simmer point	2460 °C
Intermolecular forces	unknown
Ionic radii	unknown
Nuclide	9
Main energy level	(Xe) 4f^5 6s^2
First ionization energy	534.6 kJ/mol
Second ionization energy	1050 kJ/mol
Standard potential	-2.42 V

Generally,

• Used in plasma and paint production (Fig. **1.8**),

• Used in battery production,

• Used as a heat source,

• Used in thickness measuring instruments,

• Used in portable X-ray welding,

• Used in electrical blankets [20].

Fig. (1.8). Uses of Promethium (Pm) **a)** Electric blanket **b)** Plastic paint **c)** Battery.

1.2.8. Samarium (Sm)

Samarium (Sm) is a silvery-white, bright and hard metal with a proton number of 62 and an atomic mass of 150,4 g/mol. Samarium is slowly oxidized at room temperature. The properties of Sm are summarized in Table **1.8**. It instantly blazes at 150 °C. It is electropositive and forms hydroxide gradually with hot water and rapidly with cold water. Samarium salts are dangerous to health because they are slightly toxic, leading to eye and skin irritation in the body. Samarium is never pure, but like other REEs, it is found in the structures of a variety of minerals such as monazite, bastnäsite, and samarskite [21]. Generally, it is used in:

Table 1.8. The properties of Samarium (Sm).

Proton number	62
Atomic weight	150.35 g/mol
Pauling electronegativity scale	1.2
Intensity	6.9 g/cm^3 at 20 °C
Liquefaction point	1072 °C

(Table 1.8) cont.....

Simmer point	1790 °C
Intermolecular forces	unknown
Ionic radii	unknown
Nuclide	11
Main energy level	(Xe) $4f^6\ 6s^2$
First ionization energy	542.3 kJ/mol
Second ionization energy	1066 kJ/mol
Standart potential	-2.41 V

• Permanent magnet production,

• X-ray lasers, precision-guided weapons (Fig. **1.9**),

• Carbon-based lighting,

• Ethyl alcohol hydrogenation and dehydrogenation of Samarium oxide production,

• X-ray radiology applications,

• Optical glasses due to absorption of infrared rays,

• The absorption of neutrons in nuclear power plants [15].

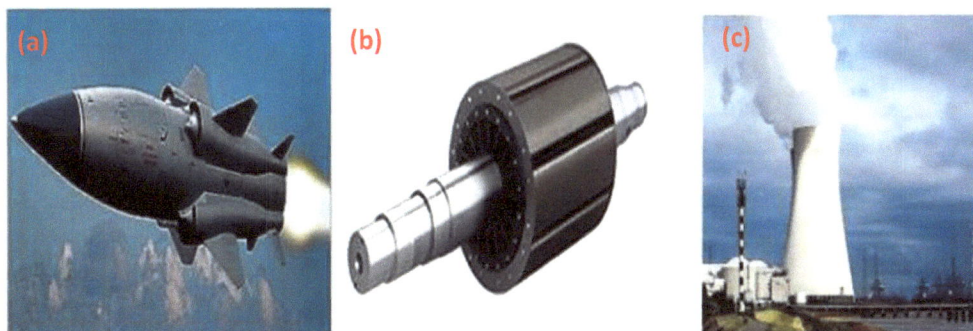

Fig. (1.9). Uses of Samarium **a)** Precision guided weapon **b)** Permanent magnet rotors **c)** Nuclear power plants.

1.2.9. Europium (Eu)

Europium (Eu) is a silvery-white metal with a proton number 63 and an atomic mass of 151.96 g/mol. Eu oxidizes rapidly in the air. It is moderate hardness. Eu is a malleable and flexible metal. Due to its crystal structure, it is similar to lead in hardness. The crystal structure is the body-centered cubic lattice. It becomes superconducting when cooled and compressed. The properties of Eu are summarized in Table **1.9**. It is very effective and ignites in the weather at temperatures between 150 °C and 180 °C. Like calcium, it reacts in water to produce europium hydroxide and hydrogen gas. Europium oxide (europium) is broadly performed as a phosphorus doping agent in TV sets and computer monitors [22]. Generally;

Table 1.9. The properties of Europium (Eu).

Proton number	· 63
Atomic weight	167.26 g/mol
Pauling electronegativity scale	1.2
Intensity	5.25 g/cm^3 at 20 °C
Liquefaction point	1522 °C
Simmer point	2510 °C
Intermolecular forces	Unknown
Ionic radii	Unknown
Nuclide	9
Main energy level	(Xe) 4f^{12} 6s^2
First ionization energy	587.6 kJ/mol
Second ionization energy	1149 kJ/mol
Standard potential	-2.30 V

• Used in lasers.

• Used to obtain red color on TVs (Fig. **1.10**).

• The phosphorous property of Europium oxide is used to separate Euro and other coins from counterfeits.

• It's used in control rods in nuclear reactors.

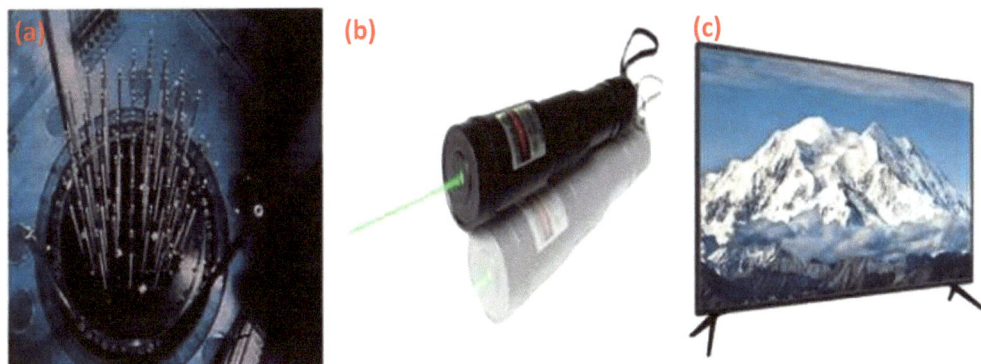

Fig. (1.10). Uses of Europium **a)** Nuclear reactors control rods **b)** Laser **c)** Tv.

1.2.10. Gadolinium (Gd)

Gadolinium (Gd), with proton number 64 and an atomic mass of 156.9 g/mol, is a silvery-white metallic element. The properties of Gd are summarized in Table **1.10**. In nature, gadolinium is not found in a pure form but in various minerals like bastnäsite and monazite. It is malleable and soft. Unlike other REEs, Gd is comparatively resolute in dry weather; however, it darkens rapidly in moist weather. Gd reacts with water slowly and dissolves in dilute acid. Gd is a ferromagnetic metal with superconducting properties. It also shows a strong magnetic property at room temperature [23].

Table 1.10. The properties of Gadolinium (Gd).

Proton number	64
Atomic weight	157.25 g/mol
Pauling electronegativity scale	1.1
Intensity	7.9 g/cm^3 at 20 °C
Liquefaction point	1313 °C
Simmer point	3266 °C
Intermolecular forces	unknown
Ionic radii	unknown
Nuclide	13
Main energy level	(Xe) 4f^7 5d^1 6s^2
First ionization energy	591.4 kJ/mol
Second ionization energy	1065.5 kJ/mol
Standard potential	-2.40 V

Generally,

• Gadolinium mixtures are used to produce phosphorus for color TVs,

• Used in reactor control rods due to the properties of several isotopes in the neutron capture area,

• Used in microwaves (Fig. **1.11**),

• Used MR imaging applications,

• Used in refractory applications,

• Used in X-ray screens,

• Used in computer chips,

• Used in tomography scintillators,

• Used in magnetic cooling devices,

• Used in ceramics.

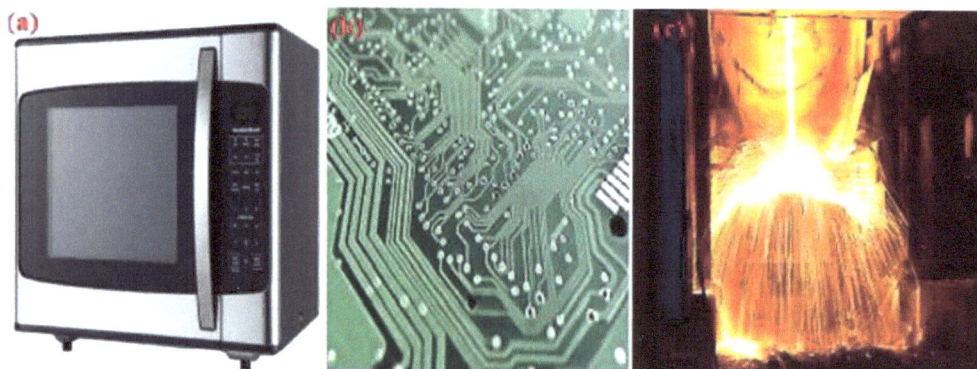

Fig. (1.11). Uses of Gadolinium (Gd) **a)** Microwave **b)** Computer chips **c)** Refractory applications.

1.3. HEAVY RARE-EARTH ELEMENTS

1.3.1. Terbium (Tb)

Terbium (Tb), with proton number 65 and an atomic mass of 159 g/mol, is a silvery grey metallic element. The properties of Tb are summarized in Table **1.11**. It is a malleable and ductile material. Tb is not found in the pure form in nature, but It is found in many minerals, together with other REEs. Tb is also easily oxidized, rare, and expensive but it is an important element among lanthanites due to its application areas. Although Tb has no environmental threat to animals or plants, Tb powder and its composition are very irritating when in contact with the eye and cutaneous. The toxicity of Tb was not studied elaborately.

Table 1.11. The properties of Terbium (Tb).

Proton number	65
Atomic weight	158.92534 g/mol
Pauling electronegativity scale	1.2
Intensity	8.3 g/cm^3 at 20 °C
Liquefaction point	1360 °C
Simmer point	3041 °C
Intermolecular forces	unknown
Ionic radii	unknown
Nuclide	9
Main energy level	(Xe) 4f^9 6s^2
First ionization energy	563.5 kJ/mol
Second ionization energy	1109.6 kJ/mol
Standart potential	-2.39 V

Generally,

• It is used in electric motors used in hybrid cars.

• The most common usage is in green phosphorus (Fig. **1.12**).

Fig. (1.12). Uses of terbium (Tb) **a)** Electric motors **b)** Green phosphorus **c)** Fluorometry devices.

• Used for very precise measurements of fluorometry devices.

• Used in measuring the characteristic wavelength of fluorescent light.

• Used in biological and medical research.

• Used in "SoundBug" speakers.

• Terbium-iron-cobalt alloys are used in CDs and DVDs [24].

1.3.2. Dysprosium (Dy)

Dysprosium (Dy), with proton number 66 and an atomic mass of 162,5 g/mol is a metallic, bright silver chemical element. The properties of Dy are summarized in Table **1.12**. In nature, Dy is seen in a natural shape; however, it's seen in many minerals, including monazite and bastnäsite. It is comparatively constant, whether at room temperature, but it is easily untwined in diluted or intensive mineral acids by hydrogen release. It is malleable sufficient to be hacked with bolt cutters and could be processed without sparks when overheating is prevented. Dysprosium is a highly sensitive ferromagnetic at below 85K. It is often used for the production of nanomagnets [25].

Table 1.12. The properties of Dysprosium (Dy).

Proton number	66
Atomic weight	162.50 g/mol
Pauling electronegativity scale	1.2
Intensity	8.6 g/cm^3 at 20 °C
Liquefaction point	1412 °C
Simmer point	3562 °C
Intermolecular forces	Unknown
Ionic radii	Unknown
Nuclide	13
Main energy level	(Xe) 4f^{10} 6s^2
First ionization energy	571.2 kJ/mol
Second ionization energy	1124 kJ/mol
Standart potential	-2.435 V

Generally,

• Used in the production of high-strength permanent magnets,

• Used for joint treatment of rheumatism,

• Used on radiation badges to detect and monitor radiation effects,

• Used in processes of chemical reactions,

• Used in laser making, together with vanadium and other REEs,

• Dysprosium-nickel mixture is used as an absorbent rod in nuclear reactors. This material has the quality of absorbing neutrons without a change in their size when exposed to prolonged neutron bombardment.

•Dysprosium sonar sensor in Terfonel-D, used in positioners (Fig. **1.13**),

• Dysprosium phosphite (DyP) is used as a semiconductor in high-power, laser diodes, and high-frequency implementations,

• Used to increase the operating temperature in electric vehicles,

• It is used in dosimeters to monitor ionizing radiation [15].

Fig. (**1.13**). Uses of dysprosium (Dy) **a)** Terfonel-D **b)** Cermet **c)** Dosimeter.

1.3.3. Holmium (Ho)

Holmium (Ho), with proton number 67 and atomic mass of 164,94 g/mol is metallic, malleable, silvery-white element. The properties of Ho are summarized in Table **1.13**. It's constant in dry weather at room temperature. It is the only element of hexagonal crystal structure among REEs. Ho has the highest magnetic permeability. Like other REEs, holmium is not pure form in nature. It occurs in rare soil minerals, especially with other minerals like gadolinite and monazite. It is a malleable and flexible metal. It is resistive and constant to corrosion in dry weather at average temperature and pressure. However, in humid weather and at higher temperatures, it is rapidly oxidized to occur a yellowish oxide. It has a silver gloss in its pure form [16].

Table 1.13. The properties of Holmium (Ho).

Proton number	67
Atomic Weight	164.9 g/mol
Pauling electronegativity scale	1.2
Intensity	8.8 g/cm^3 at 20 °C
Liquefaction point	1474 °C
Simmer point	2695 °C
Intermolecular forces	unknown
Ionic radii	unknown
Nuclide	4
Main energy level	(Xe) 4f^{11} 6s^2
First ionization energy	580 kJ/mol
Second ionization energy	1136.6 kJ/mol
Standard potential	-2.32 V

Generally,

• Due to its neutron absorption properties, it is employed in the fabrication of control rods in nuclear power plants,

• Used in laser systems to detect objects from remote distances and to create 3D images,

• Used in the production of misleading systems for helicopters and fighter jets to protect against missiles,

• Used in laser operations commonly used in medicine,

• Used in sensitive radar systems (Fig. **1.14**),

• Used to give yellow color to cubic zirconia jewelry [15].

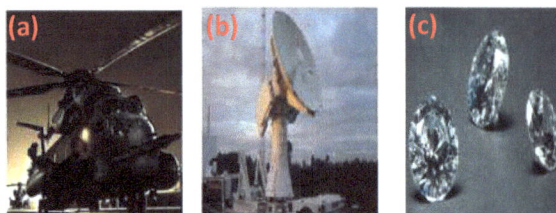

Fig. (1.14). Uses of holmium (Ho) **a)** Surface-to-air missile **b)** Precision radar system **c)** Zirconia jewelry.

1.3.4. Erbium (Er)

Erbium (Er), with proton number 68 and an atomic mass of 167,2 g/mol, is a silver metallic element. The properties of Er are summarized in Table **1.14**. Like other REEs, this element is not found pure in the earth, but it is seen in monazite sand ore. Er is a malleable, soft metal but is constant in the weather and doesn't oxidize as rapidly as many other REEs. Er gives a characteristically sharp adsorption spectrum with near-infrared, visible light, and ultraviolet. It also appears as a product of the core division. Since there are only electrons in the Erbium, it is quite paramagnetic. Due to its magnetic properties, it is an important element in fiber optic cables, laser devices, and various coloring processes. It gives pink and red-like shades to the samples. In surgical laser devices, it is a busy metal. It is a compound of lasers used to treat skin diseases.

Table 1.14. The properties of Erbium (Er).

Proton number	68
Atomic weight	167.26 g/mol
Pauling electronegativity scale	1.2
Intensity	9.2 g/cm^3 at 20 °C
Liquefaction point	1522 °C
Simmer point	2510 °C
Intermolecular forces	unknown
Ionic radii	unknown
Nuclide	9
Main energy level	(Xe) 4f^{12} 6s^2
First ionization energy	587.6 kJ/mol
Second ionization energy	1149 kJ/mol
Standard potential	-2.30 V

Generally,

• Used in fiber optic cables and medical lasers,

• In fiber optic cables, it acts as a laser amplifier,

• It is used in the nuclear industry with vanadium and titanium alloys,

• Used in glass coloring applications (Fig. **1.15**),

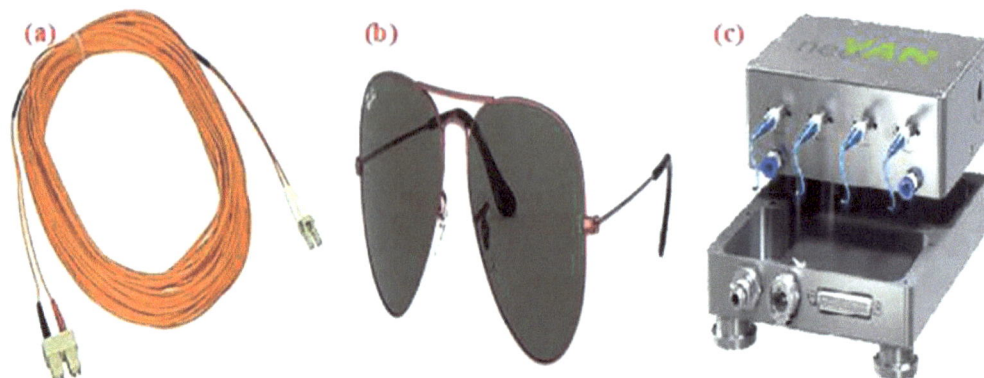

Fig. (1.15). Uses of Erbium **a)** Fiber optic cable **b)** Sunglass **c)** Artificial gem stone.

• Used in porcelain coloring,

• Used as a coloring element in similar materials such as artificial jewelry and sunglasses,

• Used for coloring oxide glasses and porcelain enamel secrets [26].

1.3.5. Thulium (Tm)

Thulium (Tm) is a brightly colored metal with atomic number 69 and an atomic weight of 168,9 g/mol. Like most REEs, it is similarly ductile, malleable, and malleable. The properties of Tm are summarized in Table **1.15**. Tm is the second RE among lanthanides after promethium, Tm reacts with water and slowly darkens in the air, but it has high corrosion resistance compared to other REEs. It should be kept away from humidity. Tm, like other lanthanides, is the element that causes poisoning, so it should be used carefully. Tm is the precious element, which limits the application areas of this element. Generally,

Table 1.15. The properties of Thulium (Tm).

Proton number	69
Atomic weight	168.93 g/mol
Pauling electronegativity scale	1.2
Intensity	9.3 g/cm^3 at 20 °C
Liquefaction point	1545 °C
Simmer point	1947 °C
Intermolecular forces	unknown

(Table 1.15) cont.....

Ionic radii	unknown
Nuclide	5
Main energy level	(Xe) $4f^{13}\ 6s^2$
First ionization energy	595.3 kJ/mol
Second ionization energy	1160.7 kJ/mol
Third ionization energy	2284 kJ/mol
Standart potential	-2.28 V

• Applied as a resource of radiation in X-ray instruments,

• Natural thulium is used in the production of ferrites in terms of magnetic materials produced from used ceramics in the production of microwave instruments (Fig. **1.16**),

• Used in portable x-ray devices and laser applications [27].

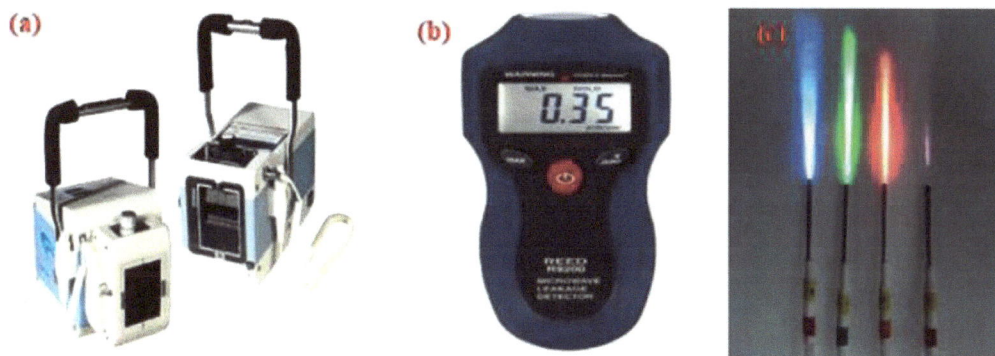

Fig. (1.16). Uses of Thulium (Tm) **a)** Portable x-ray devices **b)** Microwave **c)** Laser applications.

1.3.6. Ytterbium (Yb)

Ytterbium (Yb) is soft, silver-colored metal with proton number 70 and an atomic mass of 173,04 g/mol. The properties of Yb are summarized in Table **1.16**. It is found in gadolinite and monazite minerals. Yb is normally difficult to extract from other REEs, but the distinction between ion change and solvent extraction methods improved in the late twentieth century has been simplified. It is a malleable and very soft element. Yb reacts slowly with water and oxidizes in the air. Ytterbium has no biological mission, but it is stated that its salts excite metabolism. Yb irritates the eye and skin. All compositions should be stored in closed containers, preserved from humidity and air, and subjected to highly toxic. Metallic Y dust poses an explosion and fire hazard. Y poses no threat to animals and plants, and its salts are introduced into the chemical industry as catalysts instead of counted on as toxic and polluting.

Table 1.16. The properties of Ytterbium (Yb).

Proton number	70
Atomic weight	173.04 g/mol
Intensity	7 g/cm^3 at 20 °C
Liquefaction point	824 °C
Simmer point	1466 °C
Intermolecular forces	unknown
Nuclide	unknown
Main energy level	9
First ionization energy	(Xe) 4f^{14} 6s^2
Second ionization energy	602.4 kJ/mol
Third ionization energy	1172.3 kJ/mol
Intermolecular forces	2472.3 kJ/mol
Standard potential	-2.27 V

Generally,

• It is used as a durability agent in stainless steel,

• Used in experiments in the chemical industry,

• Used in experiments in the field of metallurgy,

• Used in diamond processing,

• Used in infrared lasers (Fig. **1.17**),

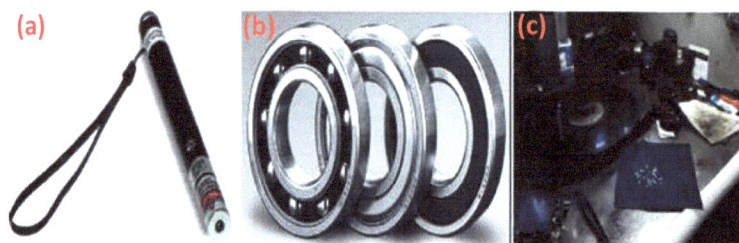

Fig. (1.17). Uses of Ytterbium (Yb) products **a)** Infrared lasers **b)** Stainless steels **c)** Diamond processing.

• Used in flame ball production,

• Used in nuclear medicine [28],

1.3.7. Lutetium (Lu)

Lutetium (Lu), with proton number 71 and an atomic mass of 174,967 g/mol, is a silver metallic element. It is the heaviest, denser, rare, and most expensive metal of REEs. The properties of Sc are summarized in Table **1.17**. It is sometimes used as a catalyst in metal alloys and various processes. Lu is a relatively stable and corrosion-resistant metal in the air. It is resistant to corrosion in the air. It is a silver-white metal in the humid air. It is easily soluble in weak acids. It burns easily at 150 degrees to form lutetium oxide. Lu is slightly toxic by swallowing, but its insoluble salts are not toxic. Like other REEs, Lu is considered to have a low toxicity rate, but this and its compositions should be used with caution. Lu's powder metal is in danger of explosion and burning but Lutetium poses no surroundings danger to animals and plants.

Table 1.17. The properties of Lutetium (Lu).

Proton number	71
Atomic weight	174.97 g/mol
Intensity	9.7 g/cm^3 at 20 °C
Liquefaction point	1663 °C
Simmer point	3395 °C
Intermolecular forces	unknown
Ionic radii	unknown
Nuclide	9
Main energy level	(Xe) $4f^{14} 5d^1 6s^2$
First ionization energy	522.7 kJ/mol
Second ionization energy	1339 kJ/mol
Standard potential	-2.25 V

Generally,

• Lutesium is used as a catalyst for many of its radioactive isotopes,

• Performed in the disintegration of petroleum products,

• Used in medicine and tomography devices,

• Used as a catalyst for some polymerization and hydrogenation studies,

• Used in detectors on positron emission tomography (Fig. **1.18**) [29, 30].

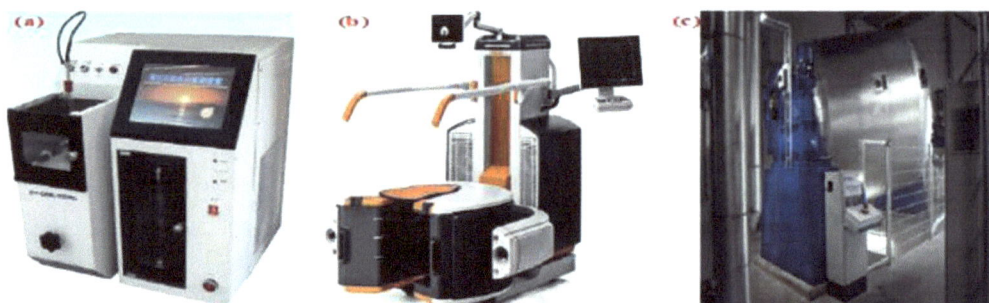

Fig. (1.18). Uses of Lutetium **a)** The disintegration of petroleum products **b)** Tomography devices **c)** Polymerization.

CONCLUSION

Due to their unique physical and chemical properties, REEs are essential components of diverse defense, energy, industrial, military technology, and low-carbon technologies. In the glass industry, REE raw materials are extensively preferred to polish glasses and used as additives that lead to the color and special properties of the glass. Petroleum refining and automotive catalytic converters need lanthanum and Cerium-based catalytic materials. Furthermore, Neodymium-iron materials, the strongest known type of magnet, are widely used in magnet applications. In addition, nickel-metal hydride batteries use anodes made of Lanthanum-based alloys. Thus, the application areas of REEs are shown variously owing to their special properties.

REFERENCES

[1]   Hurst, C.A. *China's Ace in the Hole rare-earth Elements*; Foreign military studies office (ARMY): Fort Leavenworth KS, **2010**.

[2]   Ji, X.H.; Zhang, Q.Y.; Xu, J.Q.; Zhao, Y.M. Rare-earth hexaborides nanostructures: Recent advances in materials, characterization and investigations of physical properties. *Prog. Solid State Chem.,* **2011**, *39*(2), 51-69.
        [http://dx.doi.org/10.1016/j.progsolidstchem.2011.04.001]

[3]   Tanaka, T.; Yoshimoto, J.; Ishii, M.; Bannai, E.; Kawai, S. Elastic constants of LaB_6 at room temperature. *Solid State Commun.,* **1977**, *22*(3), 203-205.
        [http://dx.doi.org/10.1016/0038-1098(77)90272-1]

[4]   Aprea, A.; Maspero, A.; Masciocchi, N.; Guagliardi, A.; Albisetti, A.F.; Giunchi, G. Nanosized rare-earth hexaborides: Low-temperature preparation and microstructural analysis. *Solid State Sci.,* **2013**, *21*, 32-36.
        [http://dx.doi.org/10.1016/j.solidstatesciences.2013.04.001]

[5]   Lihong, B.; Wurentuya, B.; Wei, W.; Yingjie, L.; Tegus, O. Chemical synthesis and microstructure of nanocrystalline RB_6 (R = Ce, Eu). *J. Alloys Compd.,* **2014**, *617*, 235-239.

[http://dx.doi.org/10.1016/j.jallcom.2014.06.207]

[6] Dou, Z.; Zhang, T.; Guo, Y.; He, J. Research on preparation optimization of nano CeB_6 powder and its high temperature stability. *J. rare-earths,* **2012,** *30*(11), 1129-1133.
[http://dx.doi.org/10.1016/S1002-0721(12)60192-6]

[7] Jiang, N.; Wang, W.M.; Fu, Z.Y.; Wang, H.; Wang, Y.C.; Zhang, J.Y. Influence of preparation parameter on the grain size of LaB_6 powder synthesized by SHS with reduction process. *Adv. Mat. Res.,* **2010,** *105*, 351-354.

[8] Xiao, Y.; Zhang, X.; Li, R.; Liu, H.; Hu, Y.; Zhang, J. Field electron emission characteristics of single-crystal GdB_6 conductive ceramics. *J. Electron. Mater.,* **2020,** *49*(9), 5622-5630.
[http://dx.doi.org/10.1007/s11664-020-08251-2]

[9] Zhou, S.; Zhang, J.; Liu, D.; Lin, Z.; Huang, Q.; Bao, L.; Ma, R.; Wei, Y. Synthesis and properties of nanostructured dense LaB_6 cathodes by arc plasma and reactive spark plasma sintering. *Acta Mater.,* **2010,** *58*(15), 4978-4985.
[http://dx.doi.org/10.1016/j.actamat.2010.05.031]

[10] Ağaoğulları, D.; Duman, İ.; Öveçoğlu, M.L. Synthesis of LaB_6 powders from La_2O_3, B_2O_3 and Mg blends *via* a mechanochemical route. *Ceram. Int.,* **2012,** *38*(8), 6203-6214.
[http://dx.doi.org/10.1016/j.ceramint.2012.04.073]

[11] Liu, H.; Zhang, X.; Ning, S.; Xiao, Y.; Zhang, J. The electronic structure and work functions of single crystal LaB_6 typical crystal surfaces. *Vacuum,* **2017,** *143*, 245-250.
[http://dx.doi.org/10.1016/j.vacuum.2017.06.029]

[12] Zhang, H.; Zhang, Q.; Tang, J.; Qin, L.C. Single-crystalline CeB_6 nanowires. *J. Am. Chem. Soc.,* **2005,** *127*(22), 8002-8003.
[http://dx.doi.org/10.1021/ja051340t] [PMID: 15926810]

[13] Iain, S.; Chassé, M. Scandium. **2016.**

[14] Lenntech, B.V. Lenntech. *Water Treat. air Purification,* **2008.**

[15] Yildiz, N. Rare Earth Elements Nadir Toprak Elementleri, **2016.**
[http://dx.doi.org/10.13140/RG.2.2.27743.87206]

[16] Housecroft, C.E. *AG Sharpe Inorganic Chemistry,* **2005,** *579*, 27.

[17] Jones, A.P.; Wall, F.; Williams, C.T. *rare-earth minerals: chemistry, origin and ore deposits*; Vol. *7*, **1995.**

[18] Eyring, L.; Gschneidner, K.A.; Lander, G.H. *Handbook on the physics and chemistry of rare-earths*; Vol. *32*, **2002.**

[19] Szabadváry, F. *Handbook of the Chemistry and Physics of the rare-earths*; Vol. *11*, **1998.**

[20] Krebs, R.E. *The history and use of our earth's chemical elements: a reference guide*; Greenwood Publishing Group, **2006.**

[21] Conradson, S.D. Los Alamos National Laboratory, Chemistry Division, Materials Science and Technology Division and Nuclear Materials Technology Division, Los Alamos, NM 87545, USA. *J. Solid State Chem.,* **2005,** *178*(2), 521-535.
[http://dx.doi.org/10.1016/j.jssc.2004.09.029]

[22] Enghag, P. *Encyclopedia of the elements: technical data-history-processing-applications*; John Wiley & Sons, **2008.**

[23] Patnaik, P. *Handbook of inorganic chemicals*; Vol. *529*, **2003.**

[24] Merhari, L. *Hybrid nanocomposites for nanotechnology*; Springer, **2009.**
[http://dx.doi.org/10.1007/978-0-387-30428-1]

[25] Cotton, F.A.; Wilkinson, G.; Murillo, C.A.; Bochmann, M.; Grimes, R. *Advanced Inorganic*

Chemistry; Vol. *6*, **1988**.

[26]　Gray, T. *Elements: A Visual Exploration of Every Known Atom in the Universe*; Black Dog & Leventhal, **2012**.

[27]　Greenwood, N.N.; Earnshaw, A. *Chemistry of the Elements*; Elsevier, **2012**.

[28]　Gmelin, L. *Hand-book of Chemistry* Cavendish Society, Vol. *1*, **1848**.

[29]　Stwertka, A. *A Guide to the Elements*; Oxford University Press, **2002**.

[30]　Chang, R. *General chemistry: the essential concepts*; McGraw-Hill: Boston, **2008**.

CHAPTER 2

The Rare-Earth Hexaborides

Abstract: Rare-earth hexaborides (REB_6) are composed of rare-earth elements and octahedral 3D boron units. In Chapter 1, rare-earth elements were examined in detail; in this part, the REB_6 will be explained. Hence, rare-earth hexaborides (REB_6) consisting of rare-earth elements and octahedral bor units are a group of ceramic materials that have a simple cubic structure with Pm3m symmetry. Their low electronic work function, low electrical resistance, and thermal expansion coefficient (in some temperature ranges), as well as high hardness and stiffness, high chemical and thermal stability, and melting points, provide a wide range of industrial uses from metallurgy to electronics.

Keywords: Nanomaterials, advanced ceramics, low work function.

2.1. INTRODUCTION

Due to the different properties of boron and its derivatives, Scientist and engineers have focused on its usage in industrial areas [1]. Turkey has 72.2% of the world's boron reserves; this is followed by Russia with 8.5% and the United States with 6.8% [2]. Boron, which has 5 atomic numbers, exhibits a semi-metallic characteristic [3]. They can be combined with almost all elements in the periodic table for the formation of the boride complex [4]. A system is required to classify a large group of boron compounds since boron can form compounds with most of the elements in the periodic table. Kiessling classified boride compounds into four main groups [1, 2].

I. Borates consist of isolated boron atoms such as M_4B and M_2B. When the percentage of boron in the compound increases, the formation of isolated pairs increases.
II. Borides consist of boron chains such as MB, and M_3B_4.
III. 3 borides of 2D boron atomic networks such as MB_2, and M_2B_5.
IV. Borides consist of 3D boron frames such as MB_4, MB_6, and MB_{12}.

The structural classification of Kiessling is shown in Table **2.1**. There are also other classification methods for borides, such as the content of boron in the comp-

Mikail Aslan and Cengiz Bozada

ound and the location of the metallic section in the periodic table [1]. The classification of binary metal borides can also be done by referring to the main metal group of the periodic table [3, 4].

Table 2.1. The classification of borides done by Kiesling [2].

Kiessling Group	Atomic Ratio	Examples
Isolated B atoms	M_4B, M_3B, M_2B	Mn_4B, Be_2B, Ni_3B
Pairs of B atoms	M_3B	V_3B_2
Single Chains	MB	FeB, NiB
Branched Chains	$M_{11}B_9$	$Ru_{11}B_9$
Doubled Chains	M_3B_4	Ta_3B_4, Cr_3B_4
Layer Networks	MB_2	$TiB_2, MgB_2, YB_2, ReB_2$
3D Frameworks	MB_4, MB_6, MB_{12}	$CaB_6,$ ZrB_{12}, YB_{12}

This group of REB_6 has a wide range of industrial uses from metallurgy to electronics because of their low electronic work function, low electrical resistance, and thermal expansion coefficient (in some temperature ranges), as well as high hardness and stiffness, high chemical and thermal stability and melting points. These properties are due to the octahedron units providing strong covalent bondings. These bonds allow the REB_6 to exhibit superior properties [2].

The applications and developments of REB_6 nanostructures attract a lot of attention. Examining the mechanical properties of REB_6 structures is extremely important in widening their applications. Nonlinear effects are known to be important in nanostructural materials. The nonlinearity elastic properties are very important in defining nonlinearity influences in mechanical behavior. Therefore, nonlinear influences must work in the flexibility of REB_6 [3].

REB_6 are used in various materials, such as warplanes, bicycles, oil refineries, fiber optic cables, high-resolution cameras, telescopes, night vision binoculars, qualified camera lenses, fiber optic cables, artificial gemstones, lasers, magnets, electric vehicle engines, headphones, enamel colorings, computer chips, sunglasses, and tomography equipment.

In this part, we have focused on the chemical and physical properties and application areas of each REB_6 structure.

2.2. RARE-EARTH HEXABORIDES

2.2.1. Lanthanum Hexaboride (LaB$_6$)

Lanthanum hexaboride (LaB$_6$) is a vacuum-stable resistant advanced ceramic substance with a melting point of 2210 °C, unsolvable in H$_2$O and HCl acid. The stoichiometric samples are intensely purple-violet, whereas those rich in boron ($\geq$LaB$_{6.07}$) are blue [5]. The main usage areas of LaB$_6$ are; as a coated material in cathodes, as an optical MEMS-based sensor in the NIR range, in radar systems, in electronic and infrared applications, and environmental protection, as refining and synthetic organic chemicals. Due to the low evaporation rate concerning high-temperature electron emission and low work function [6], its application is wide. LaB$_6$ cathodes are used as devices and techniques such as X-ray tubes, electron lithography, electron microscopes, electron beam source, free-electron lasers, and microwave tubes (Fig. **2.1.**).

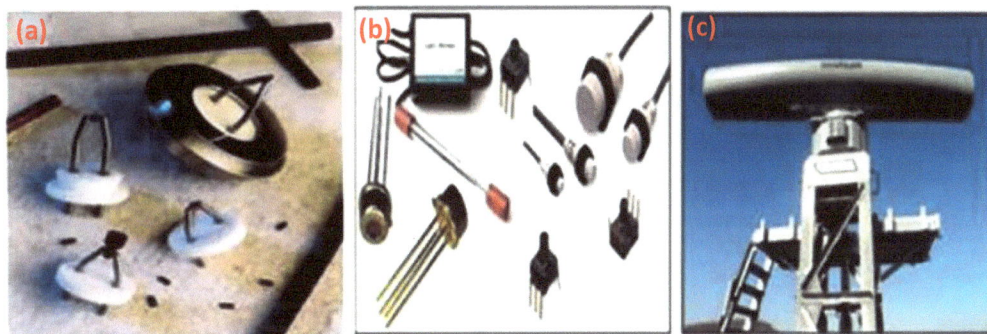

Fig. (2.1). Uses of LaB$_6$ **a)** LaB$_6$ cathode **b)** Optical MEMS-based sensor **c)** Radar system.

2.2.2. Cerium Hexaboride (CeB$_6$)

Among the rare-earth hexaboride family, cerium hexaboride (CeB$_6$) as an electron resource is appropriate to be used as thermionic electron emitters because it has a lower operating function, a lower operating temperature, and a higher electrical resistance. Recently, some investigators have focused on the field emission properties of one-dimensional nanoelectronic CeB$_6$, as well as nanomachine applications. It also has resistance to cathode poisoning. For this reason, it can be used in free-electron lasers, electron lithography, microwave tubes, X-ray tubes, electron beam welding, and electron microscopes [7]. Some of the applications are given in Fig. (**2.2**).

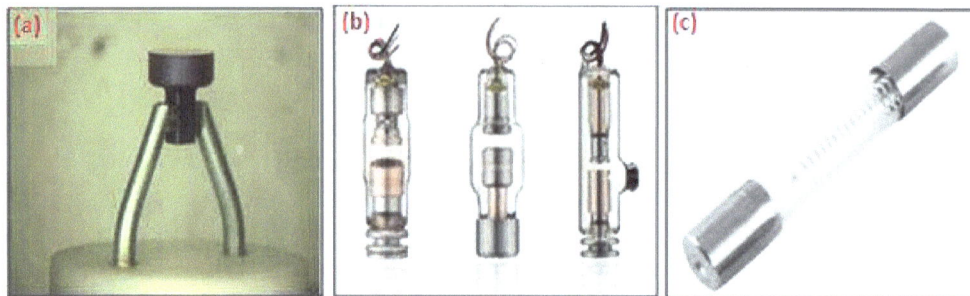

Fig. (2.2). Uses of CeB_6 **a)** CeB_6 cathode **b)** X-ray tube **c)** Microwave tube.

2.2.3. Praseodymium Hexaboride (PrB_6)

Praseodymium hexaboride (PrB_6) has an acceptable low operating function (3.12 eV) and lower volatility. Furthermore, PrB_6 have a lower ambient temperature and longer usage time compared to traditional cathodic emission substance [8]. PrB_6 materials are used in the application of nuclear technologies, high-resolution detectors, electrical coating of resistors, high-energy optical systems, and thermionic materials (Fig. **2.3**) due to their properties, such as strength, high chemical stability, high neutron absorption, as well as and low-temperature coefficient, stable specific resistance and low electronic operating function [9].

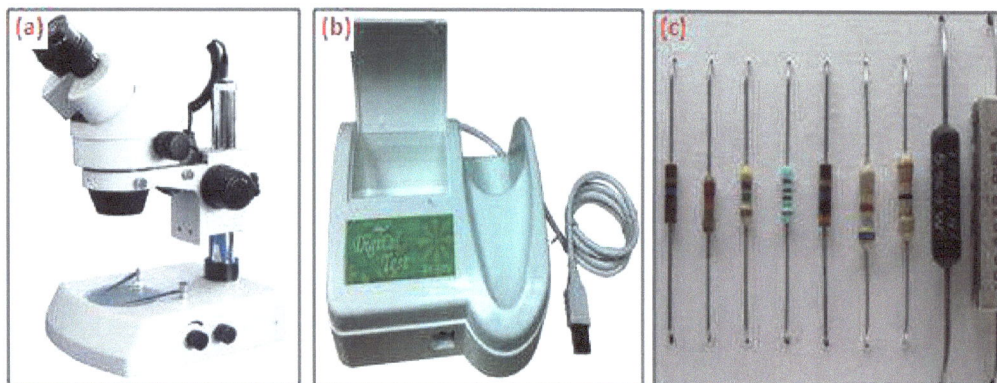

Fig. (2.3). Uses of PrB_6 **(a)** High energy optical system **(b)** High-resolution detector **(c)** Electrical coatings of resistors.

2.2.4. Neodymium Hexaboride (NdB_6)

Neodymium hexaboride (NdB_6) has a high melting point, high thermal conductivity, low thermal expansion coefficient, and an acceptable hardness (hardness at 900 °C is 1010 kg / mm^2) [10]. In addition, NdB_6 is used as a thermionic electron source since it has a low work function (1.6 eV). NdB_6 is also a good material for neutron absorption applications in a nuclear reactor. The amount of cross-sectional area for neutron absorption is high (3800 barns) [11]. Some of the application areas are given in Fig. (**2.4**).

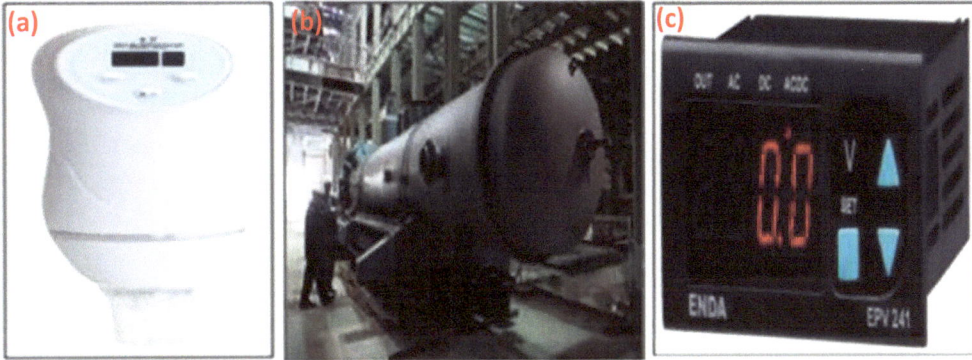

Fig. (2.4). a) Vacuum electronic device **b)** Neutron absorber application in nuclear reactor **c)** thermionic electron source.

2.2.5. Europium Hexaboride (EuB₆)

Europium hexaboride (EuB_6) has very important properties. The nuclear capabilities of the EuB_6 are remarkable. Especially for fast and high thermal neutrons, both boron (B) and europium (Eu) atoms have very high neutron absorption [12]. It is also used in EuB_6 cinema projectors and carbon arc lamps (Fig. **2.5**) [13].

Fig. (2.5). Uses of EuB_6 **(a)** Thermionic emission material **(b)** Cinema projector **(c)** Carbon arc lamp.

2.2.6. Samarium Hexaborides (SmB₆)

Samarium Hexaborides (SmB_6) is defined as a material having high rigidity, high chemical and thermal stability, and low thermal expansion coefficient. SmB_6 is a typical intermediate valence material that has an energy gap close to the Fermi level. Due to the low operating function, SmB_6 indicates excellent optical properties that can be used in thermionic emission, field-induced emission and thermal applications [14]. SmB_6 is also a topological kondo insulator and exhibits heavy fermion behavior [15].

SmB_6 is also used as a control bar, shielding, and hardware in nuclear applications, as shown in Fig. (**2.6**).

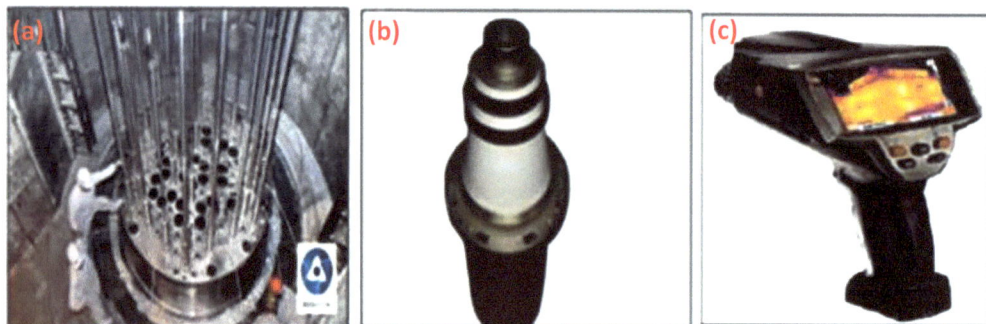

Fig. (2.6). Uses of SmB_6 (**a**) Control bar in nuclear applications (**b**) Thermionic emission (**c**) Thermal site-induced emissions.

2.2.7. Gadolinium Hexaboride (GdB$_6$)

Gadolinium hexaboride (GdB$_6$) has low volatility, low electrical resistance, low spray coefficient, high mechanical strength, and chemical resistance [16]. Due to its high NIR (near-infrared) absorption, GdB_6 is used as a potential solar radiation protective material (Fig. **2.7**). It is also suitable as a thermionic electron-emitting material [17].

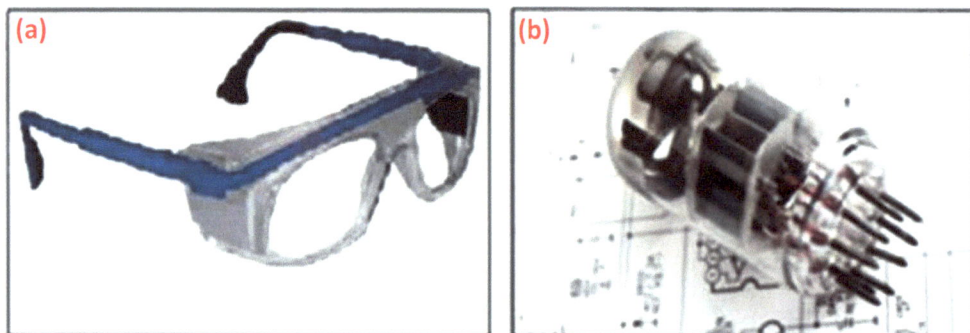

Fig. (2.7). Uses of GdB$_6$ (**a**) Solar radiation shielding material (**b**) Thermionic electron-emitting material.

2.2.8. Erbium Hexaboride (ErB$_6$)

Erbium hexaborür (ErB$_6$) is a rare-earth hexaboride compound. It is one of the main compounds formed in reactions between erbium and boron [18]. Due to the natural structure of hexaborides in rare-earth and the strong interaction of boron octahedral in the crystal, these compounds show a high degree of lattice matching [19]. It is used as a suitable extinguishing agent for flammable materials (Fig. **2.8**).

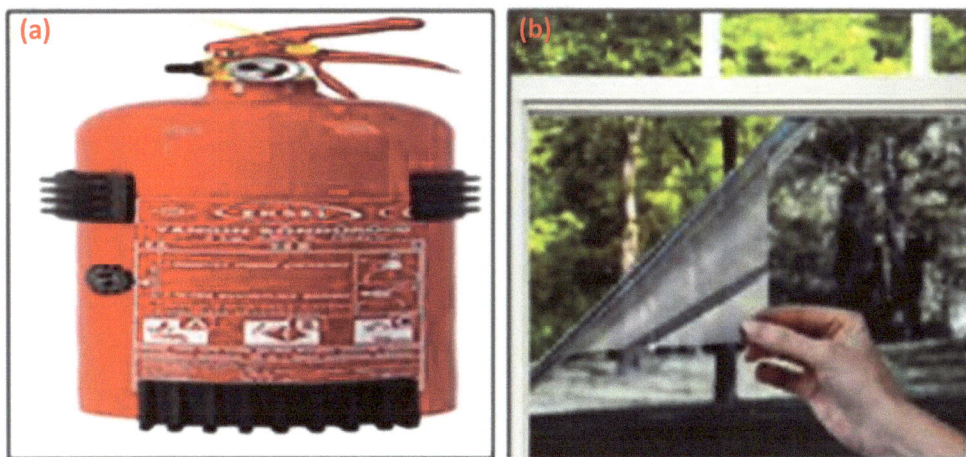

Fig. (2.8). Uses of ErB$_6$ **(a)** Firefighting material **(b)** Solar protective window film.

2.2.9. Ytterbium Hexaboride (YbB$_6$)

Ytterbium hexaboride (YbB$_6$) has a semiconductor with a bandgap of approximately 0.3 eV among the Yb 5d conduction states and B 2p valence states. YbB$_6$ is an insulator but not a kondo insulator [20]. It is a promising candidate for applications based on topological surface conditions (Fig. **2.9**) [21]. YbB$_6$ is also diamagnetic at T$_c$ = 12,5 K [22].

Fig. (2.9). Uses of YbB$_6$ **(a)** Solar energy system **(b)** Thin layer (film) **(c)** Heat radiation shielding material.

2.2.10. Scandium Hexaboride (ScB$_6$)

The impossibility of obtaining Scandium Hexaboride (ScB$_6$) is because of dimensional factors, as well as very powerful reactions of Sc with B [23]. The boron atoms in the crystal lattice are shifted from equal positions to the closed positions of the cubic base cell surfaces. An electron beam generation cathode is made from a rare-earth metal hexaboride such as ScB$_6$ (Fig. **2.10**).

Fig. (2.10). Electron beam generation cathode.

2.2.11. Thulium Hexaboride (TmB$_6$)

Thulium Hexaboride (TmB$_6$) indicates antiferromagnetic properties [24]. TmB$_6$ is used in polymeric sheets and rigid sheets to make laminated glass. The work function shows a positive temperature coefficient in the temperature range 1100-1800 K. Rapid evaporation was shown at 1800 K. The molecular Weight of TmB$_6$ is 233.796 [25]. Some applications of TmB$_6$ are given in Fig. (**2.11**).

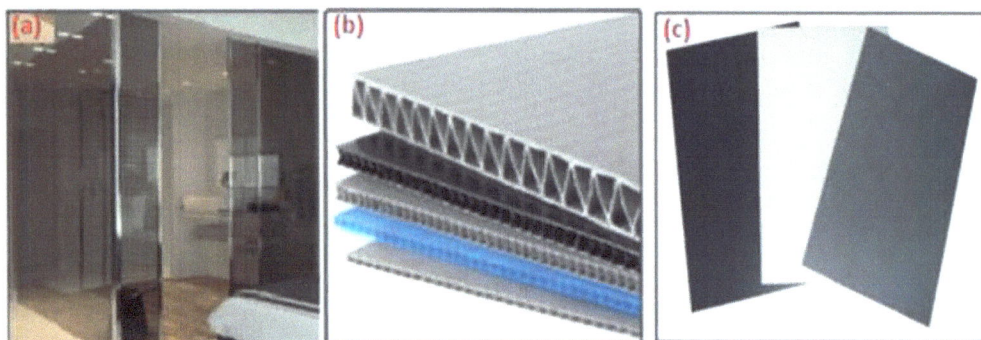

Fig. (2.11). Uses of TmB$_6$ **(a)** Laminated glass **(b)** Polymeric sheet **(c)** Rigid sheet.

2.2.12. Dysprosium Hexaboride (DyB$_6$)

Dysprosium hexaboride (DyB$_6$) is a heavy rare-earth hexaboride. DyB$_6$ has an incompatible melting property, which makes it difficult to obtain large single crystals [26]. DyB$_6$ is an antiferromagnet at T_N = 26 K. T_N is defined as Néel temperature or magnetic regulation temperature [27]. DyB$_6$ is used for marking metal and plastic parts, heatwave protection components, and laser radiation (Fig. **2.12**).

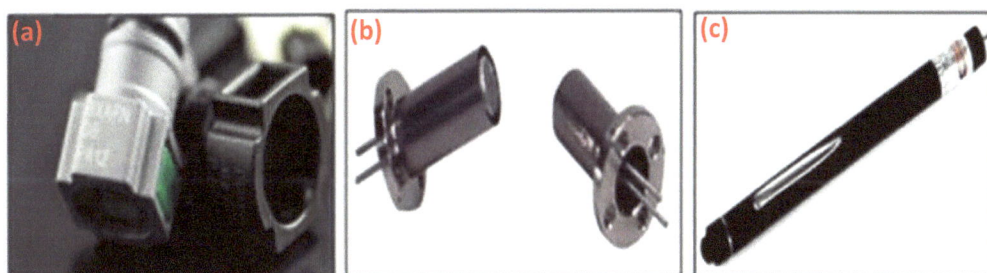

Fig. (2.12). Uses of DyB$_6$ **(a)** Marking of metal and plastic parts **(b)** Heatwave protection component **(c)** Diode laser radiation pen.

2.2.13. Yttrium Hexaboride (YB$_6$)

YB$_6$ is nonmagnetic and a superconductor. The superconductor passing temperature is high at T_c = 7.5 K [28]. YB$_6$ has the primitive cubic CaB$_6$ type of crystal structure with a lattice parameter of ~4.154 Å [29]. YB$_6$ shows quite high nano-hardness (24.6 GPa). The Burgers vectors of dislocations have been identified as <100> for YB$_6$ [30]. Some applications of YB$_6$ are given in Fig. (**2.13**).

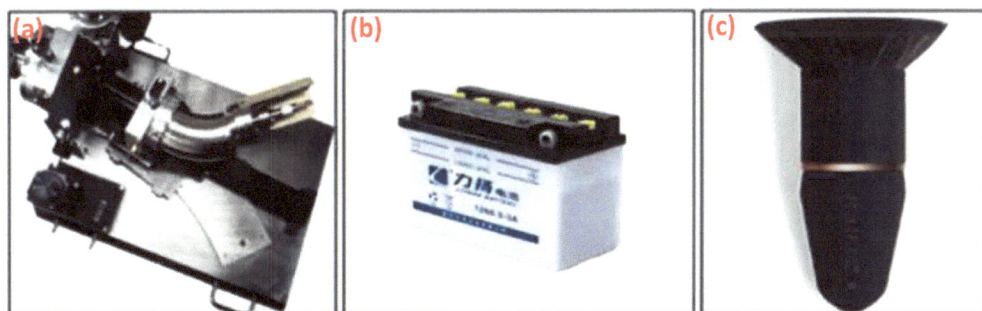

Fig. (2.13). Uses of YB$_6$ **(a)** Soft x-ray monochromator **(b)** Motorcycle battery **(c)** Powder buffing brush.

2.2.14. Holmium Hexaboride (HoB$_6$)

HoB$_6$ with Γ_5 triplet ground state exhibits an antiferromagnetic (AFM) ordering at T_N=5.6K and a ferroquadrupole (FQ) ordering at T_Q=6.1 K and neutron scattering experiments show the crystal structure of HoB$_6$ changes from cubic to trigonal at T_Q [31]. Some applications of HoB$_6$ are given in Fig. (**2.14**).

Fig. (2.14).   Uses of HoB$_6$ (**a**) Additive manufacturing (**b**) Heat radiation shielding (**c**) Infrared radiation.

2.2.15. Terbium Hexaboride (TbB$_6$)

TbB$_6$ are antiferromagnetic at magnetic transition temperatures (T_N) = 19.5 [32]. The antiferromagnetic state of TbB$_6$ shows a tetragonal symmetry [33]. Some applications of TbB$_6$ are given in Fig. (**2.15**).

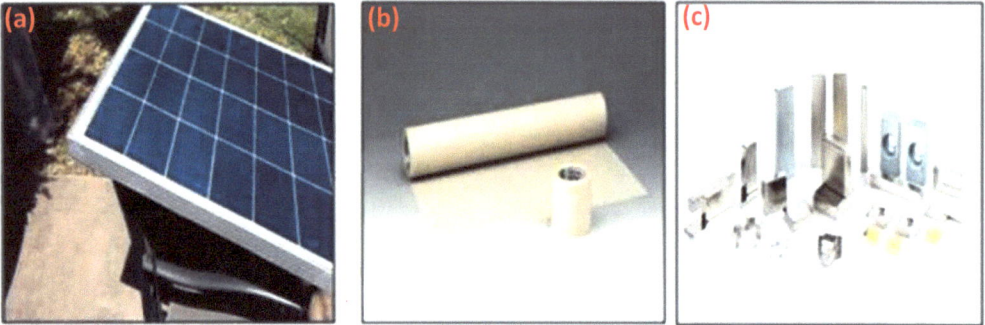

Fig. (2.15).   Uses of TbB$_6$ (**a**) Solar panels (**b**) Fluororesin film (**c**) Nd-Fe-B permanent magnetic material.

CONCLUSION

REB$_6$ advanced materials show superior properties, such as high mechanical strength, high melting point, low vaporization, high mechanical strength, high chemical inertness, and long-time usage. Especially, examining the mechanical properties of REB$_6$ structures is tremendously significant for enlarging their applications. Thus, REB$_6$ materials were used in a lot of components of industrial sectors.

REFERENCES

[1] Abazova, A.K.H.; Kushkhov, K.H.B.; Mukozheva, R.A.; Tlenkopachev, M.R.; Uzdenova, A.S.; Vindizheva, M.K. Electrolytic method of producing ultrafine cerium hexaboride powder using cerium source and boron source. *Russ. Pat. No. RU2466090-C1,* **2012**.

[2] Gökçe, H.; Ağaoğulları, D.; Öveçoğlu, M.L.; Duman, İ.; Boyraz, T. Characterization of microstructural and thermal properties of steatite/cordierite ceramics prepared by using natural raw materials. *J. Eur. Ceram. Soc.,* **2011**, *31*(14), 2741-2747.
[http://dx.doi.org/10.1016/j.jeurceramsoc.2010.12.007]

[3] Bakr, M.; Kinjo, R.; Choi, Y.W.; Omer, M.; Yoshida, K.; Ueda, S.; Takasaki, M.; Ishida, K.; Kimura, N.; Sonobe, T.; Kii, T.; Masuda, K.; Ohgaki, H.; Zen, H. Back bombardment for dispenser and lanthanum hexaboride cathodes. *Phys. Rev. Spec. Top. Accel. Beams,* **2011**, *14*(6), 060708.
[http://dx.doi.org/10.1103/PhysRevSTAB.14.060708]

[4] Ağaoğullari, D.; Balci, Ö.; Gökçe, H.; Duman, İ.; Öveçoğlu, M.L. Synthesis of magnesium borates by mechanically activated annealing. *Metall. Mater. Trans., A Phys. Metall. Mater. Sci.,* **2012**, *43*(7), 2520-2533.
[http://dx.doi.org/10.1007/s11661-012-1109-5]

[5] Lundstrom, T. Structure, defects and properties of some refractory borides. *Pure Appl. Chem.,* **1985**, *57*(10), 1383-1390.
[http://dx.doi.org/10.1351/pac198557101383]

[6] Yamamoto, N.; Rokuta, E.; Hasegawa, Y.; Nagao, T.; Trenary, M.; Oshima, C.; Otani, S. Oxygen adsorption sites on the PrB_6 (100) and LaB_6 (100) surfaces. *Surf. Sci.,* **1996**, *348*(1-2), 133-142.
[http://dx.doi.org/10.1016/0039-6028(95)00989-2]

[7] Bao, L-H.; Zhang, J-X.; Zhou, S-L.; Zhang, N.; Xu, H. Floating zone growth and thermionic emission property of single crystal CeB_6. *Chin. Phys. Lett.,* **2011**, *28*(8), 088101.
[http://dx.doi.org/10.1088/0256-307X/28/8/088101]

[8] Takeda, H.; Kuno, H.; Adachi, K. Solar control dispersions and coatings with rare-earth hexaboride nanoparticles. *J. Am. Ceram. Soc.,* **2008**, *91*(9), 2897-2902.
[http://dx.doi.org/10.1111/j.1551-2916.2008.02512.x]

[9] Xu, J.; Chen, X.; Zhao, Y.; Zou, C.; Ding, Q. Single-crystalline PrB_6 nanowires and their field-emission properties. *Nanotechnology,* **2007**, *18*(11), 115621.
[http://dx.doi.org/10.1088/0957-4484/18/11/115621]

[10] Otani, S.; Nakagawa, H.; Nishi, Y.; Kieda, N. Floating zone growth and high temperature hardness of rare-earth hexaboride crystals: LaB_6, CeB_6, PrB_6, NdB_6, and SmB_6. *J. Solid State Chem.,* **2000**, *154*(1), 238-241.
[http://dx.doi.org/10.1006/jssc.2000.8842]

[11] Sonber, J.K.; Murthy, T.S.R.; Sairam, K.; Bedse, R.D.; Hubli, R.C.; Suri, A.K. Synthesis and characterization of neodymium hexaboride powder. Proceedings of the DAE-BRNS fourth interdisciplinary symposium on materials chemistry, **2012**.

[12] Xu, J.; Chen, X.; Zhao, Y.; Zou, C.; Ding, Q.; Jian, J. Self-catalyst growth of EuB_6 nanowires and nanotubes. *J. Cryst. Growth,* **2007**, *303*(2), 466-471.
[http://dx.doi.org/10.1016/j.jcrysgro.2006.12.040]

[13] Swift, T.; Moore, M.; Rose-Glowacki, H.; Sanchez, E. The Economic Benefits of the North American rare-earths Industry. *rare-earth Technol. Alliance,* **2014**, *1* Vol. *1*, **2014**.

[14] Ağaoğulları, D.; Balcı, Ö.; Öveçoğlu, M.L.; Suryanarayana, C.; Duman, İ. Synthesis of bulk nanocrystalline samarium hexaboride. *J. Eur. Ceram. Soc.,* **2015**, *35*(15), 4121-4136.
[http://dx.doi.org/10.1016/j.jeurceramsoc.2015.07.037]

[15] Hatnean, M.C.; Lees, M.R.; Paul, D.M.; Balakrishnan, G. Large, high quality single-crystals of the new Topological Kondo Insulator, SmB_6. *Sci. Rep.,* **2013**, *3*(1), 3071.

[http://dx.doi.org/10.1038/srep03071] [PMID: 24166216]

[16]   Ali, N. Anomalous electrical and magnetic properties of gadolinium hexaboride. *J. Appl. Phys.,* **1988**, *63*(8), 3583-3585.
[http://dx.doi.org/10.1063/1.340701]

[17]   Han, W.; Wang, Z.; Li, Q.; Liu, H.; Fan, Q.; Dong, Y.; Kuang, Q.; Zhao, Y. Autoclave growth, magnetic, and optical properties of GdB_6 nanowires. *J. Solid State Chem.,* **2017**, *256*, 53-59.
[http://dx.doi.org/10.1016/j.jssc.2017.08.026]

[18]   Samsonov, G. V. High-temperature compounds of rare-earth metals with nonmetals. **1965**.

[19]   Schmidt, P.H.; Joy, D.C. Low work function electron emitter hexaborides. *J. Vac. Sci. Technol.,* **1978**, *15*(6), 1809-1810.
[http://dx.doi.org/10.1116/1.569847]

[20]   Kang, C.J.; Denlinger, J.D.; Allen, J.W.; Min, C.H.; Reinert, F.; Kang, B.Y.; Cho, B.K.; Kang, J.S.; Shim, J.H.; Min, B.I. Electronic structure of YbB_6 : Is it a topological insulator or not? *Phys. Rev. Lett.,* **2016**, *116*(11), 116401.
[http://dx.doi.org/10.1103/PhysRevLett.116.116401] [PMID: 27035312]

[21]   Xia, M. Angle-resolved photoemission spectroscopy study on the surface states of the correlated topological insulator YbB_6. *Sci. Rep.,* **2014**, *4*(1), 1-6.

[22]   Tarascon, J.M.; Etourneau, J.; Dordor, P.; Hagenmuller, P.; Kasaya, M.; Coey, J.M.D. Magnetic and transport properties of pure and carbon-doped divalent RE hexaboride single crystals. *J. Appl. Phys.,* **1980**, *51*(1), 574-577.
[http://dx.doi.org/10.1063/1.327309]

[23]   Barantseva, I.G.; Paderno, Y.B. Possibility of existence of scandium hexaboride. *Soviet Powder Metallurgy and Metal Ceramics,* **1981**, *20*(9), 635-638.
[http://dx.doi.org/10.1007/BF00801573]

[24]   Etourneau, J. Critical survey of rare-earth borides: Occurrence, crystal chemistry and physical properties. *J. Less Common Met.,* **1985**, *110*(1-2), 267-281.
[http://dx.doi.org/10.1016/0022-5088(85)90333-9]

[25]   Raleigh, H.D. *Direct Energy Conversion Devices: A Literature Search*; United States Atomic Energy Commission, Division of Technical Information, **1961**.
[http://dx.doi.org/10.2172/4032358]

[26]   Kunii, S.; Iwashita, K.; Matsumura, T.; Segawa, K. Mechanism of the magnetization process of DyB_6. *Physica B,* **1993**, *186-188*, 646-648.
[http://dx.doi.org/10.1016/0921-4526(93)90662-P]

[27]   Granovsky, S.A.; Markosyan, A.S. Large crystal structure distortion in DyB_6 studied by X-ray diffraction. *J. Magn. Magn. Mater.,* **2003**, *258-259*, 529-531.
[http://dx.doi.org/10.1016/S0304-8853(02)01133-2]

[28]   Kasuya, T. Exchange-pair Jahn-Teller effects in GdB_6. *J. Magn. Magn. Mater.,* **1997**, *174*(1-2), L28-L32.
[http://dx.doi.org/10.1016/S0304-8853(97)00229-1]

[29]   Kunii, S.; Kasuya, T.; Kadowaki, K.; Date, M.; Woods, S.B. Electron tunneling into superconducting YB_6. *Solid State Commun.,* **1984**, *52*(7), 659-661.
[http://dx.doi.org/10.1016/0038-1098(84)90728-2]

[30]   Sekido, N.; Ohmura, T.; Perepezko, J.H. Mechanical properties and dislocation character of YB4 and YB_6. *Intermetallics,* **2017**, *89*, 86-91.
[http://dx.doi.org/10.1016/j.intermet.2017.05.024]

[31]   Yamaguchi, T. Thermal expansion and ultrasonic measurements of ferroquadrupole ordering in HoB_6. *Physica B,* **2003**, *329-333*, 622-623.

[http://dx.doi.org/10.1016/S0921-4526(02)02524-3]

[32]   Granovsky, S.A.; Amara, M.; Galéra, R.M.; Kunii, S. Magnetic and magneto-elastic properties of a single crystal of TbB$_6$. *J. Phys. Condens. Matter,* **2001**, *13*(29), 6307-6321.
[http://dx.doi.org/10.1088/0953-8984/13/29/304]

[33]   Amara, M.; Galéra, R.M.; Aviani, I.; Givord, F. Macroscopic and microscopic investigation of the antiferromagnetic phase of TbB$_6$. *Phys. Rev. B Condens. Matter Mater. Phys.,* **2010**, *82*(22), 224411.
[http://dx.doi.org/10.1103/PhysRevB.82.224411]

CHAPTER 3

The Structures of Rare-Earth Hexaborides

Abstract: The structures of rare-earth hexaborides can be nanoparticles, nanowires, nanotubes, nanorods, nano-obelisks, nanocubes, nanocrystals and nanocons. These types of structures indicate superior properties, such as excellent mechanical, electronic, and optical properties. For these reasons, they are used in thermionic materials, electrical coating for resistors, sensors, and high-energy optical systems. Furthermore, their low work functions make them special for the design of optical devices, such as a cathode substance for cold (field) emission

Keywords: Low work function, thermionic materials, nanostructures.

3.1. INTRODUCTION

Nano-sized materials are divided into different classes, such as nanoparticles, nanowires, nanotubes, nanorods, nano-obelisks, nanocubes, nanocrystals, and nanocons. The main step for new developments in nanotechnology includes the design, functional use, and production of nanostructured materials and tools in the production of nanoparticles. Nanosized materials always have exceptional mechanical, electronic, or optical properties. The optical features of nanomaterials are very significant to investigate because of their presence of surface plasmon resonance character and nanoscale dimension [1]. REB_6 nanostructures, both in the shape of nanorods, nanocubes, nanowires, nanoparticles, nanotubes, nano-obelisk, nanoblets, nanoawls, amorphous, nanocrystals, and nanocones, have drawn significant notice because of their extensive diversity of possible implementations in thermionic materials, electrical coating for resistors, sensors and high energy optical system. REB_6 nanostructures are considered the best thermionic electron source for field emissions applications due to their high melting point, high chemical resistance, high conductivity, low volatility at high temperatures, and low work functions. REB_6 are accepted as productive cathodes for improved vacuum electronic tools [2]. Having a low work function is a very important criterion in terms of being designed as a cathode substance for cold (field) emissions that offer more than a hundred times brightness. Hence, REB_6 is the most excellent thermionic electron resource [3].

Mikail Aslan and Cengiz Bozada

Many methods, such as spark plasma sintering (SPS), chemical vapor deposition (CVD), and mechanochemical, floating zone methods, have been used to produce REB_6 nanostructures. SPS is an important technique for REB_6 nanostructures at low temperatures. The SPS technique is one of the suitable methods to produce nanostructured intense REB_6 with superb properties [4]. One of the methods used to develop REB_6 nanostructures is chemical vapor deposition. REB_6 nanostructures are potentially used as point electron emitters for applications including cold emission, Edison effect (thermionic emission), and thermal field-induced electron emission for TEM, SEM, smooth panel screens, and other electronic tools that need high-performance electrons [5].

One of the methods used to produce high-purity REB_6 nanostructures is the mechanochemical method. The mechanochemical method is an important method for REB_6 nanostructures to show very good properties [6]. One of the methods used for REB_6 nanostructures to have excellent applications possibilities as thermionic cathode substances is the floating zone method. Moreover, this method usually provides a contamination phase that directly reduces emission and crystal quality characteristics [7]. The different structure types of REB_6 are listed in Table **3.1**.

In this part, we have focused on different types of REB_6 structures, such as nanowires, nanotubes, nanorods and nanocubes. Furthermore, recent developments and new trends have been discussed.

3.2. NANOSTRUCTURES

3.2.1. Nanowires

Nanowires are nanostructures in a cylindrical form quite similar to carbon nanotubes. Generally, nanowires have a thickness or diameter of tens of nanometers. There are several types of nanowire insulators (SiO_2, TiO_2), semiconductors (Si, Ge, InP, GaN), metallic (Ni, Pt, Au, Fe), and carbon nanotubes. In the production of nanowires, many common laboratory techniques, such as extraction, chemical deposition, vapor transport (deposition), and vapor-liquid-solid magnification, are used [8]. Nanowires have very important applications in optoelectronics (light-interacting electronic devices), electronics, tips for bio-molecular nano-sensors, nano-electromechanical devices, advanced composites, nano-scale quantitative instruments for metallic interconnections, and as field emitters [9]. Zou *et al.* synthesized CeB_6 nanowires by the self-catalyst method. Their results indicate that CeB_6 nanowires have a diameter of approximately 20-100 nm, and the longness recumbents to a few micrometers [10]. Fig. (**3.1**) illustrates the images of the CeB_6 dendritic crystal. The diameters are approximately 30 nm, and the trunk has a longness of approximately 4 mm. It is also observed that the diameter of the branches is about 30 nm.

Table 3.1. The structures of the given materials.

Material	Structure	Fabrication Method	References
CeB_6	Nanowires	Self-catalyzed CVD	[10]
NdB_6	Nanowires	Self-catalyzed CVD	[11]
LaB_6	Nanowires	Self-catalyzed CVD	[12]
GdB_6	Nanowires	Self-catalyzed CVD	[13]
SmB_6	Nanowires	Self-catalyzed CVD	[14]
EuB_6	Nanowires	Self-catalyzed CVD	[15]
LaB_6	Nanotubes	Self-catalyzed CVD	[12]
EuB_6	Nanotubes	Self-catalyzed CVD	[15]
LaB_6	Nanorods	Aluminum flux method	[18]
LaB_6	Nanocubes	Molten salt technique	[20]
LaB_6	Nanocubes	Solid-state technique	[21]
CeB_6	Nanocubes	Electrochemical synthesis	[22]
LaB_6	Nano-obelisk	Metal-catalyzed CVD	[24]
NdB_6	Nano-obelisk	Metal-catalyzed CVD	[25]
NdB_6	Nanoparticle	Mechano-chemical alloying	[28]
SmB_6	Nanoparticle	Solid-state technique	[29]
NdB_6	Nanoparticle	Melt spinning technology	[30]
PrB_6	Nanoparticle	Metal-catalyzed CVD	[2]
SmB_6	Nanobelts	Metal-catalyzed CVD	[32]
SmB_6	Nanobelts	Metal-catalyzed CVD	[33]
$La_xPr_{1-x}B_6$	Nanoawls	Simple flux-controlled technique	[34]
EuB_6	Amorphous	Liquid plasma technique	[36]
LaB_6	Amorphous	Solid-state technique	[37]
SmB_6	Nanocrystals	Mechanochemical synthesis	[29]
PrB_6	Nanocrystals	Solid-state technique	[40]
SmB_6	Nanocrystals	Metal-catalyzed CVD	[41 - 43]

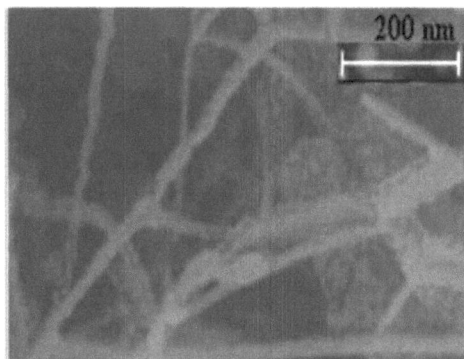

Fig. (3.1). SEM picture of a CeB_6 dendritic crystal Adopted from [10] (Copyright © 2006 Published by Elsevier B.V.).

The elaborated microstructure and morphology properties of CeB_6 nanowires are shown in Fig. (**3.2**). Also, the high-resolution transmission electron microscopy (HRTEM) lattice picture of the CeB_6 nanowire is shown in the figure. The lattice in the HRTEM is shown as cubic, and the ranges of the nearest points are measured as 0.23 nm, which corresponds to the cubic CeB_6 plane (111) plane. The diameter of the high-growth TEM picture of CeB_6 nanowires is approximately 80 nm. In Fig. (**3.2b**), the normal shape of the nanowires is visible. At the same time, it appears that there are no particles at the end of the nanowires. The electron diffraction model received throughout the (224) crystal region axis is shown in Fig. (**3.2c**). The electron diffraction pattern indicates that the CeB_6 nanowires assume a single crystal structure, which has a primitive cubic structure with the space group of Pm3m.

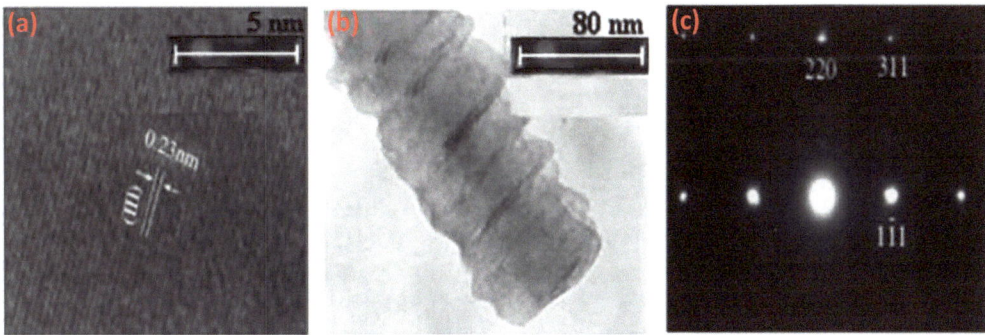

Fig. (3.2). (**a**) HRTEM of CeB_6 nanowire. (**b**) The TEM of the CeB_6 nanowire; (**c**) The typical SAED of a CeB_6 nanowire received throughout the [224], direction Adopted from [10] (Copyright ©2006 Published by Elsevier B.V.).

In another study, Ding *et al.* synthesized NdB_6 nanowires by a self-catalyst technique with Nd powders and boron trichloride (BCl_3) gas mixed with argon and hydrogen. NdB_6 nanowires have a diameter of approximately 80 nm, and the longness recumbents to a few micrometers. NdB_6 nanowires are cubic single crystals according to the TEM [11]. In another study, Xu *et al.* successfully produced LaB_6 nanowires with a self-catalyst technique by employing La powders and BCl_3 gas stirred with argon and hydrogen. Their results indicated that LaB_6 nanowires are several micrometers long and 20-200 nm thick. TEM exhibits that the nanowires are extremely crystalline [12].

In another study, Zhang *et al.* produced GdB_6 nanowires of well-characterized morphology by the CVD process. The GdB_6 nanowires have longness recumbents to more than several microns and have a rectangular cross-section with a thickness of approximately 50 nm. The GdB_6 nanowires have walls and smooth tips, which are every-ended by the {100} lattice planes. A GdB_6 single nanowire field emitter synthesized an emission current of more than 150 nA at a mean implemented area below 3.2 V/μm. The work function of GdB_6 nanowires is smaller than LaB_6 nanowire emitters, and they were found as 1.5 eV. GdB_6 nanowires have huge potential use as point electron emitters in implementations, including ensuring point emission resources for electronic tools [13].

In another study conducted by Xu *et al.*, SmB_6 nanowires were synthesized by a catalysis-free technique. It was found that the SmB_6 nanowires are single crystals with the (100) magnification direction [14]. In a similar study, Xu *et al.* effectively synthesized EuB_6 nanowires with a self-catalyst technique that used Eu powders and BCl_3 gas mixed with argon and hydrogen. Their results indicate that EuB_6 nanowires are 1-10 mm long thick and 60-300 nm. TEM analyses exhibit that the EuB_6 nanowires are crystal with (100) magnification direction [15].

3.2.2. Nanotubes

Nanotubes are nanostructures with very high mechanical and electronic properties. Nanotubes can be semiconductors or metals, depending on their structure type. In addition, it has good thermal conductivity and good flexibility, and it is a very robust, and extremely enduring material. Also, it is stable at very high temperatures and is diamagnetic, and has excellent chemical stability. Energy-dispersive X-ray spectroscopy (EDX) analyses were performed to explore the growth mechanism of REB_6 nanotubes. Xu *et al.* produced LaB_6 nanotubes by a self-catalyst technique using La powders and BCl_3 gas assorted with H and Ar. Their results reveal that LaB_6 nanowires are 20-200 nm thick and a few micrometers long [12]. The selected area electron diffraction (SAED) model is shown in Fig. (**3.3a**), where the nanotubes are polycrystalline structured material types. This is the reverse of that of nanowires, where a mono structure is considered. One of the superb properties of LaB_6 is that it is resistant to electron bombardment, so it is hard to connect the polycrystal structure of LaB_6 nanotubes to electron beam damage. The diffraction model of the nanotube agrees with the X-ray diffraction measurement result of the products obtained on the clean Si substrate, and also the SAED result of the nanorods with the lattice parameter, *a* of the primary cubic structure (100), (111), (311) can be easily indexed. The TEM image of the LaB_6 nanotubes is shown in Fig. (**3.3b**). Two non-uniform worms typed nanotubes were observed.

It is seen that the diameter of the larger nanotube is between 200-400 nm, the wall thickness is approximately 60 nm, there is a length of up to 1 μm, and the diameter of the small one is about 200 nm, and the length is about 400 nm wall width around 30 nm. In addition, the inner wall of the nanotubes is jagged and therefore has two similar nanotubes morphology.

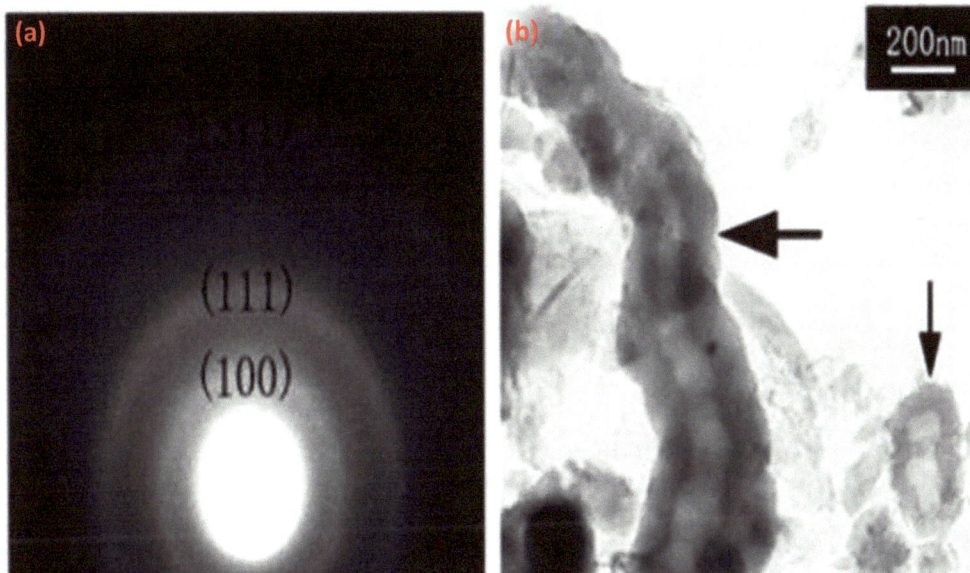

Fig. (3.3). (a) SAED model of the nanotube. The diffraction circles are arranged according to the LaB$_6$ crystal structure. **(b)** TEM of LaB$_6$ nanotubes. Two nanotubes are indicated with two arrows. Adopted from [12] (Copyright © 2006 Elsevier B.V.).

Bukatova *et al.* synthesized GdB$_6$ nanotubes in the temperature range 973-1023 K [16]. Xu *et al.* effectively synthesized EuB$_6$ nanotubes with self catalyst technique. It was found that EuB$_6$ nanotubes were polycrystalline. Moreover, they characterized the EuB$_6$ nanotubes by TEM. The nanotube has a diameter of about 100 nm and a longness of 20-30 nm [15].

3.2.3. Nanorods

In nanotechnology, nanorods are 1D dimensional structures that provide a directed path for electrical transport and are used to control the bandgap by varying the radius of rods and using the quantum size effect. Each of its dimensions is between 1-100 nm. Metals or semiconductor materials can be used for their production. The standard aspect ratios (length to width ratio) are 3-5. Nanorods are produced directly by chemical synthesis. A potential application of the nanorods is in display technologies since the reflection directions of the rods can be changed by replacing them with an applied electric field. Another application is for microelectromechanical systems (MEMS). Nanorods, together with other essential metal nanoparticles, also can be used as diagnostic agents (combining diagnostic information with drugs to improve the efficacy and safety of cancer treatments). Nanorods based on semiconductor materials have been investigated in applications, such as energy collection and light-emitting devices [17]. Jha *et al.* studied LaB_6 nanorods. The nanorods of LaB_6 have excellent field emission properties since LaB_6 nanorods exhibit low turn-on fields and high field enhancement factors.

Fig. (**3.4**) illustrates the field behavior applied against the current density of LaB_6 nanorods. The LaB_6 nanorods (20 nm×120 nm) exhibit a threshold field (4.9 $\mu A/cm^2$) and low turn-on fields of electron emission (1 $\mu A/cm^2$) [18].

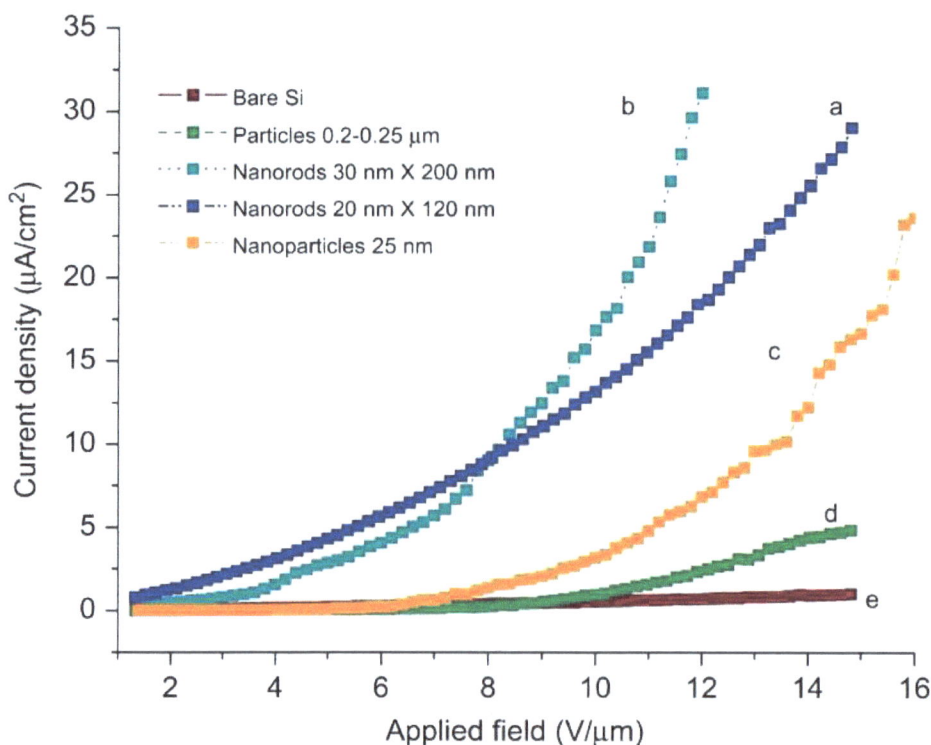

Fig. (3.4). Field emission study of LaB_6 synthesized employing polycrystal LaB_6 **(a)** Nanorod is 120 nm × 20 nm in diameter **(b)** Nanorod is 200 nm×30 nm in diameter **(c)** Nanoparticles are 25 nm in diameter **(d)** Particles arranging 200-250 nm and **(e)** Bare Si. Adopted from [18] (©2012 Elsevier Ltd.).

3.2.4. Nanocubes

As the size of the nanocubes increases, the resonance wavelengths of the high-grade plasmonic modes are separated and become apparent. The high crystalline, high purity, and single-phase properties of REB_6 nanotubes are verified by XRD, EDX, and Raman. FESEM and TEM. Aprea *et al.* developed REB_6 nanocubes using MgB_2 and hydrated rare-earth trichloride. The diameter of REB_6 nanocubes is a few nanometers [19]. In a different study conducted by Yu *et al.*, LaB_6 nanocubes were synthesized by the Ionic liquid (molten salt) technique. The microstructures of LaB_6 are shown in Fig. (**3.5**), indicating the LaB_6 particles in the specimen have superior cubes of 60-100 nm [20].

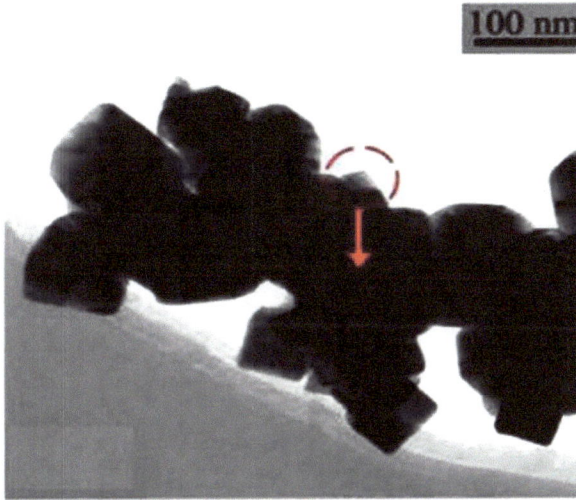

Fig. (3.5). Low-magnification TEM image of LaB_6 nanocubes Adopted from [20] (Copyright © 2018 Elsevier B.V.).

Zhang *et al.* successfully synthesized LaB_6 nanocubes having a mean particle dimension of 30 nm, using Mg, $LaCl_3$, and $NaBH_4$ in an autoclave at 400 °C. LaB_6 nanocubes with an average dimension of ~200 nm can be produced at 500 °C. The XRD could be arranged as cubic LaB_6 with the lattice parameter of α=4.154 Å for LaB_6 nanocubes [21]. Kushkhov *et al.* produced CeB_6 nanotubes by potentiostatic electrolysis of molten KCl-NaCl, including $CeCl_3$ and KBF_4. The electrolysis on the tungsten electrode was used in the range of −2.4 to −2.8 V comparatively with the quasi-stationary glassy-carbon electrode [22].

3.2.5. Nano-Obelisk

The nano-obelisk plays a significant role in minimizing a built-in electric field, enhancing light extraction, and extinguishing dislocations [23]. Brewer *et al.* fabricated LaB_6 nano-obelisks *via* different sizes and pyramidal tip heights by a metal-catalyzed CVD technique. The LaB_6 nano-obelisk was characterized by four-sided shafts with a broader basis *via* their pyramidal rectangular tips. LaB_6 nano-obelisk of dissimilar determined cone angles, pyramidal peak highness, and shaft wideness could be favorably acquired *via* the cautious position of the subgrades into 2 cm lengthy at 920-945 °C enlargement region downstream by the La precursor. The LaB_6 nano-obelisks were investigated by EDX and TEM [24]. Fig. (**3.6**) shows classic TEM images of the LaB_6 obelisk tip. The HRTEM and SAED explain the single crystal of the LaB_6 nano-obelisk *via* showed a lattice interval of 4.15 Å (Fig. **3.6**).

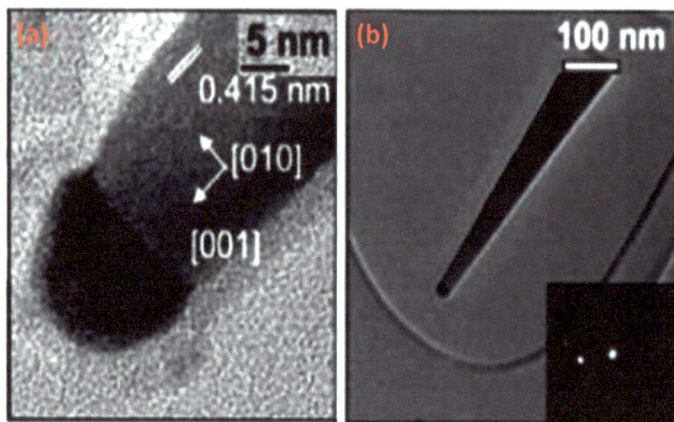

Fig. (3.6). TEM images of a LaB_6 nano-obelisk at (**a**) high and (**b**) low resolution Adopted from [24] (Copyright © 2007 American Chemical Society).

Wang *et al.* investigated a palladium-nanoparticle-catalyzed CVD approach to produce NdB_6 nano-obelisks, but the boron precursor ($B_{10}H_{14}$) is rigid and isn't simple to check its flow under CVD. Moreover, that process needs exterior warming tools and complicated duct lines to the reactor [25].

3.2.6. Nanoparticles

The nanoparticle size of REB_6 changes the plasmonic absorbance band and the atomic vibrations on the surface. The optical absorption of REB_6 nanoparticles is important in terms of the coexistence of high visible light transmission and powerful near-infrared adsorption. Methods used in the production of nanoparticles were classified from top to bottom. Examining from the bottom-up, particle formation is carried out by enlarging atomic or molecular structures through chemical reactions. Spray pyrolysis, sol-gel, chemical vapor deposition (CVD), CVD coating, and gas condensation technique are examples of this examination. The top-down examination is based upon the separation of the material into small pieces that can go down to nano-size as a result of chemical processes or mechanical energy being applied to the material from the outside. Abrasion and mechanical grinding are examples of this approach.

The optical properties of nanoparticles of LaB_6, with its high thermal stability and incredible hardness, employ it as a perfect option involved in composites and alloys for solar windows. If the size of the LaB_6 nanoparticles drops to ~2.5 nm, there is a transition to superior energy for total B_6 cluster vibration modes because of the enlarged surface area with more B_6 numbers on the surface (Fig. **3.7**) [26].

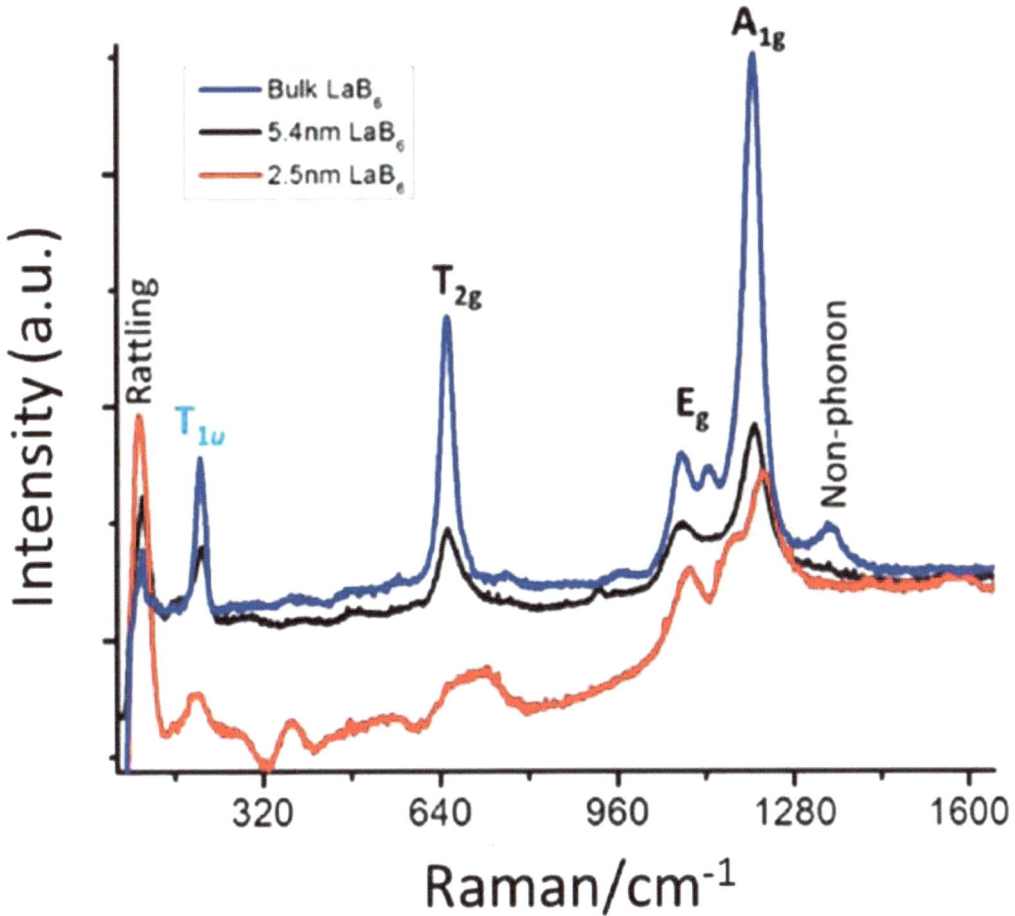

Fig. (3.7). Vibrational mode changes with changing nanoparticle sizes Adopted from [26] (Copyright © 2018 by the authors. Licensee MDPI, Basel, Switzerland).

CeB_6 has similar free-electron plasmon energies as LaB_6. The structure and shape of nanocrystalline metals play significant roles in identifying the position, number, and density of surface plasmon resonance modes. CeB_6 nanoparticles exhibit feeble annihilation in the visible interval and comparatively powerful annihilation in NIR [27]. Şimşek *et al.* fabricated NdB_6 nanoparticles by mechano-chemical alloying. The pure NdB_6 nanocrystals have an 18.24 nm crystallite size. They showed the irregular morphology of NdB_6 nanoparticle structures.

Fig. (**3.8**) indicates the TEM images of the depurated NdB_6. Fig. (**3.8a**) shows HRTEM images; nanoparticles are almost spherical in size of approximately $12 \pm$ 5 nm. The HRTEM of a single nanoparticle with a size of approximately 10.13 nm reveals obvious lattice eaves of which the d interval is nearly 0.286 nm, subtending the (101) plane of the cubic NdB_6 nanocrystalline. The low-magnification image (Fig. **3.8b**) shows that the material consists of small nanoparticles that tend to lump due to their strong inter-particle strength [28].

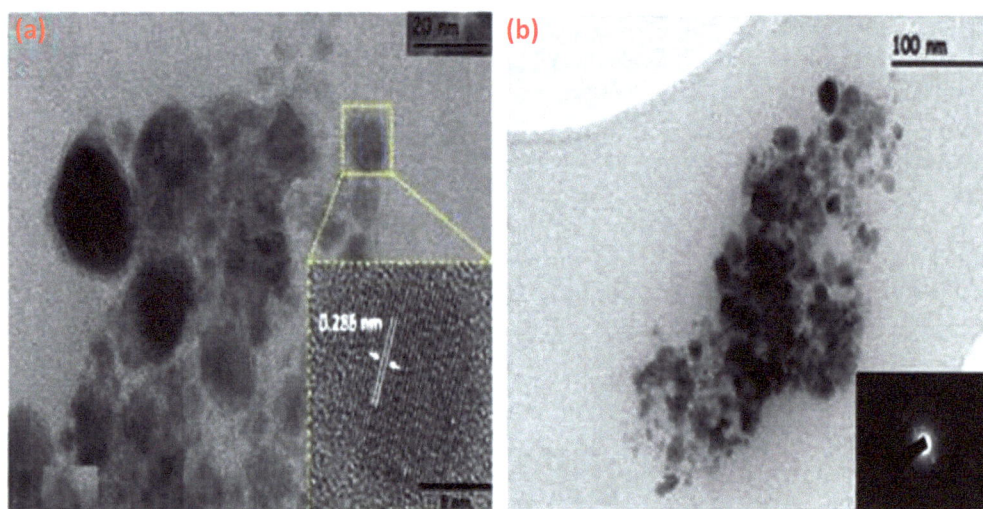

Fig. (3.8). (a) HRTEM picture of the NdB_6 nanocrystalline **(b)** Low-enlargement TEM picture of the NdB_6 nanocrystalline with SAED model Adopted from [28] (Copyright © 2019 Elsevier B.V.).

Bao *et al.* produced SmB_6 at the temperature interval of 1000-1200 °C with the Solid State reaction of $Sm_2O_3/SmCl_3$ with $NaBH_4$. The FESEM showed that the SmB_6 nanoparticles have a super crystalline cubic structure (Fig. **3.9**). Bao *et al.* found that the SmB_6 crystals were produced by $SmCl_3$ with a perfect crystallinity and more straight dispersion than Sm_2O_3 [29].

Ding *et al.* synthesized NdB_6 by melt spinning technology. The hardness, melt point, chemical, and thermostability of NdB_6 are superior, therefore, the NdB_6 particles could be employed to strengthen and refine the aluminum alloy. The SEM picture of the NdB_6 nanoparticles taken by NdB_6/Al stripes shows in Fig. (**3.10**). Most particles are in the range of 40-60 nm, and in a few quantities, the particle is bigger than 100 nm [30].

Fig. (3.9). FESEM pictures of SmB$_6$ were arranged for 2 hours at 1200 °C using SmCl$_3$ as raw material, adopted from [29] (Copyright © 2015 Elsevier B.V.).

Fig. (3.10). The NdB$_6$ particles taken from NdB$_6$/Al stripes Adopted from [30] (Copyright© 2019 Elsevier B.V.).

Chi *et al.* produced PrB$_6$ with a CVD process employing Pr powders and BCl$_3$ as initial samples. Their results indicated that the as-synthesized nanoparticles were polycrystalline with ~2-10 μm in length and 10-100 nm in diameter [2]. The TEM of a typical nanowire is illustrated in Fig. (**3.11a**). The calculated lattice parameter is nearly 0.296 nm and 0.407 nm. The angle of 45° between the two region axes was calculated. The wire-like nanoparticles with a comparatively smooth size dispersion and a longness of about 2-10 μm deposited on the subgrade could be observed in Fig. (**3.11b**).

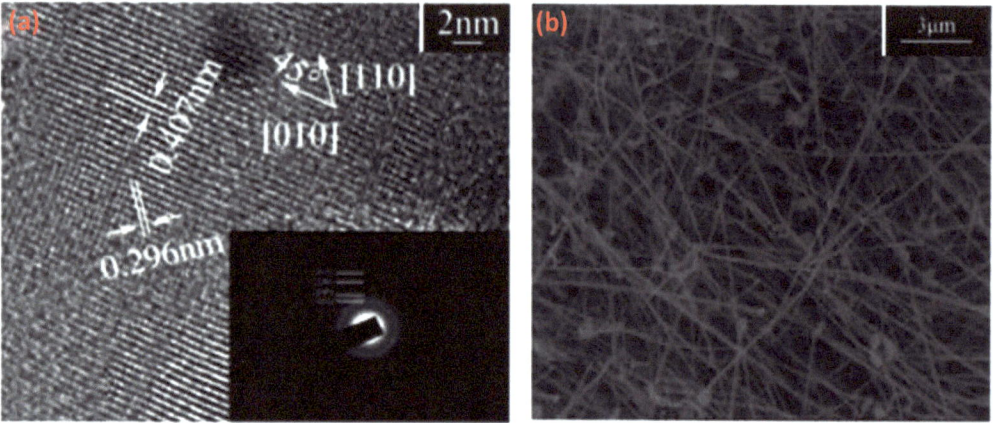

Fig.3.11. (a) HRTEM picture of a nanowire **(b)** A typical SEM image of PrB_6 nanowires Adopted from [2] Copyright © 2014 Published by Elsevier Ltd.

3.2.7. Nanobelts

Nanobelts are an important substance due to their superb physical properties, unique morphologies, and shape-induced electronic and optic properties, providing wide potential technological applications [31]. Zhang *et al.* fabricated SmB_6 nanobelt film on the Si substrate by chemical vapor deposition (CVD). SmB_6 nanobelt is cubic single crystals and their growth directions are 110. The field emission (FE) results indicate that SmB_6 nanobelts have a turn-on field of 3.24 V/µm. SmB_6 nanobelt films using B_2H_6 (diborane) and H_2 as the reaction gas source were prepared and investigated for the variable temperature field emission properties [32]. In another study, Gan *et al.* synthesized SmB_6 nanobelt film by CVD. SmB_6 nanobelt diameter and thickness were found as 40 and 500 nm, respectively. The growth directions are (110), and the length is 130 µm [33]. As shown in Fig. (**3.12**), nanobelts sequences were distributed on the substrate and were based on the substrate at an average angle of 70°. Their enlargement intensity is above 10^5 cm^{-2}. In addition, it was found that the nanobelts have an average longness of 130 µm and an average thickness of several micrometers, and their width varies from 40 to 500 nm, found the activation energy of SmB_6 nanobelt was calculated as 4.4 meV.

3.2.8. Nanoawls

Nanoawles are a significant material due to their good morphologies and superb physical properties. Han *et al.* synthesized Lanthanum-praseodymium hexaboride ($La_xPr_{1-x}B_6$) nanoawles using La powders, Pr powders, and boron trichloride

(BCl_3) gas with a simple flux-controlled technique. SEM images demonstrate that the nanoawles are a size between about 50-300 nm at the ends, a longness of 2-10 μm, and 10-80 nm at the ends.TEM results show that the nanoawles are single crystals with an enlargement direction throughout (110) [34]. Fig. (**3.13**) shows typical SEM images of $La_xPr_{1-x}B_6$ nanoawles obtained on Si substrates at 1050 °C. Fig. (**3.13**) indicates the nanoawles with a homogeneous morphology. Nanoawls is a flat-shaped substance about 2-10 μm long. $La_xPr_{1-x}B_6$ is a continuous solid solution that can be obtained nanoawles of different x values.

Fig. (3.12). The part picture of SmB_6 nanobelts. Adopted from [33] (Copyright © 2019, American Chemical Society).

Fig. (3.13). SEM images of the $La_xPr_{1-x}B_6$ acquired at 1050 °C. Adopted from [34] (Copyright © 2016 Elsevier Ltd.).

3.2.9. Amorphous

Amorphous solid atoms do not have a stable crystal structure [35]. Wei *et al.* investigated amorphous EuB_6 with the reaction between B_2H_6 and $EuCl_3$ in the existence of fluid plasma in ionic fluid surroundings. The sample illustrates an instant, long-term, and quite-selecting answer to formaldehyde at room temperature *via* a determination limit of 50 ppb. There is a powerful interaction between electron-enriched Eu regions on the EuB_6 surface and formaldehyde [36].

As seen in Fig. (**3.14a**), TEM indicates EuB_6 has an erratic morphology with a mean particle size of ~6 nm. The SAED pattern (inset) verifies the amorphous structure. In the XRD of Fig. (**3.14b**), at room temperature, there exists a peak, seen at ~30.1° (2θ). This means weakly crystalline material or amorphous. Ivashchenko *et al.* examined the thermodynamic and mechanical properties and electronic and phonon structures of amorphous LaB_6 (a-LaB_6 and c-LaB_6) by DFT. The thermal conductivities wane as it passes from a-LaB_6 to c-LaB_6. The peak at 100 cm^{-1} of the phonon spectrum of c-LaB_6 is significantly narrower than that of a-LaB_6 [37].

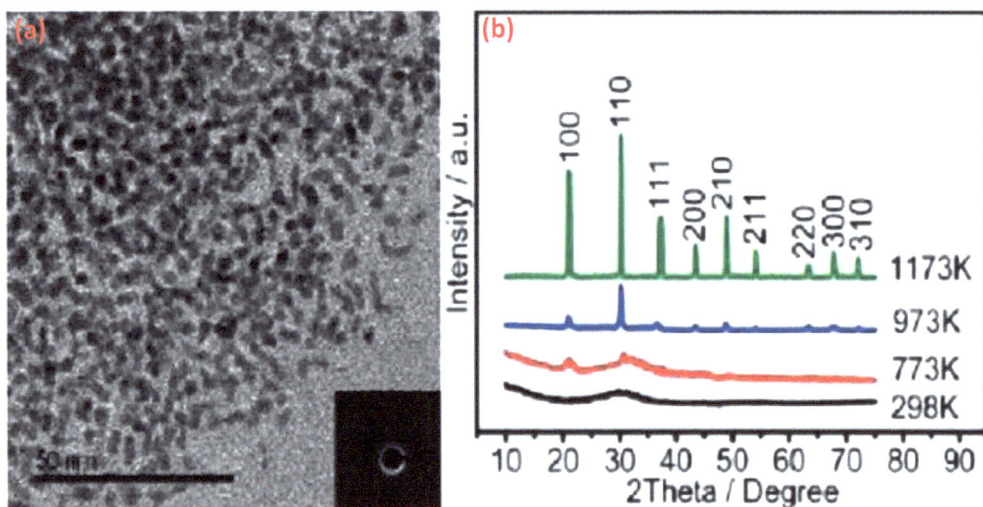

Fig. (3.14). (a) TEM image **(b)** XRD patterns of EuB_6 material. Adopted from [36] (Copyright© 2018 American Chemical Society).

3.2.10. Nanocrystals

A nanocrystalline is a particle of atoms in a single or multi-crystalline order, less than 100 nanometers in size. Nanocrystals are remarkable due to their electrical resistance, high coefficient of thermal expansion, strong strength, and low thermal conductivities [38]. Bao *et al.* fabricated SmB_6 nanocrystals *via* the solid-state method of $NaBH_4$ with $Sm_2O_3/SmCl_3$ in the temperature interval of 1000-1200°C. HRTEM shows that SmB_6 nanocrystalline has a cubic structure with high crystallinity [29]. Fig. (**3.15**) depicts a classic TEM of SmB_6 nanoparticles prepared at 1100 °C. They found that the mean grain dimension is around 40-50 nm, which is consistent with the SEM observation.

Fig. (3.15). TEM analyses of SmB_6 nanocrystals. Adopted from [29] (Copyright © 2015 Elsevier B.V.).

In another study, Chen *et al.* investigated CeB_6, PrB_6, SmB_6, EuB_6, and LaB_6 nanocrystals. The nanocrystals have a high melting point and high crystallinity. The average sizes of CeB_6, EuB_6, and LaB_6 nanocrystals are 300 nm, 200 nm, and 1 mm, respectively [39]. Öveçoğlu *et al.* synthesized SmB_6 nanocrystals from Mg, B_2O_3, and Sm_2O_3 powder blends at room temperature. SmB_6 nanocrystals are kondo insulators or heavy electron (fermion) semiconductors. Furthermore, the electrical resistivity (211,00 μΩ), density (4.91 g/cm³), and microhardness (10.08 GPa) of SmB_6 nanocrystals were found [40]. Wei *et al.* synthesized PrB_6 nanocrystals (100~300 nm) by a solid-state reaction. PrB_6 nanocrystals are the CsCl structure type. SAED patterns explain that the nanocrystals have single-crystalline nature. The reaction temperature has a huge influence on the crystal dimension, morphology, and homogeneity of the nanocrystal distribution [41].

3.2.11. Nanocone

The lengths of nanocones are several micrometers. The diameter of nanocones is also several nanometers. The nanocone is a single crystalline structure [42]. Yang *et al.* produced SmB_6 nanocone by CVD technique. SmB_6 nanocones have good crystallinity and good emission stability. The field emission (FE) of SmB_6 nanocone was studied. SmB_6 nanocones possess perfect FE behaviors. SmB_6 nanocones have an electric field low turn-on of 1.84 V/μm. The average length of SmB_6 nanocones is 1-3 μm [43].

CONCLUSION

The developments in nanotechnology have been recently going towards the design, and functional use of nanostructured materials and tools in the production of nanoparticles. Materials under the nanoscale level indicate exceptional mechanical, electronic, and optical properties. REB_6 nanostructures consisting of nanorods, nanocubes, nanowires, nanoparticles, nanotubes, nano-obelisk, nanobelts, nanoawls, amorphous, nanocrystals, and nanocone have drawn attention because of their extensive diversity of possible implementations in thermionic materials, electrical coating for resistors, sensors and high energy optical system.

REFERENCES

[1]   Karak, N. Fundamentals of Nanomaterials and Polymer Nanocomposites. In: *Nanomaterials and Polymer Nanocomposites*; Elsevier, **2019**; pp. 1-45.

[2]   Chi, M.; Zhao, Y.; Fan, Q.; Han, W. The synthesis of PrB_6 nanowires and nanotubes by the self-catalyzed method. *Ceram. Int.,* **2014**, *40*(6), 8921-8924.
        [http://dx.doi.org/10.1016/j.ceramint.2014.01.046]

[3]   Zhang, Q.Y.; Xu, J.Q.; Zhao, Y.M.; Ji, X.H.; Lau, S.P. Fabrication of Large-Scale Single-Crystalline PrB_6 Nanorods and Their Temperature-Dependent Electron Field Emission. *Adv. Funct. Mater.,* **2009**, *19*(5), 742-747.
        [http://dx.doi.org/10.1002/adfm.200801248]

[4]   Zhou, S.; Zhang, J.; Liu, D.; Lin, Z.; Huang, Q.; Bao, L.; Ma, R.; Wei, Y. Synthesis and properties of nanostructured dense LaB_6 cathodes by arc plasma and reactive spark plasma sintering. *Acta Mater.,* **2010**, *58*(15), 4978-4985.
        [http://dx.doi.org/10.1016/j.actamat.2010.05.031]

[5]   Zhang, H.; Zhang, Q.; Tang, J.; Qin, L.C. Single-crystalline CeB_6 nanowires. *J. Am. Chem. Soc.,* **2005**, *127*(22), 8002-8003.
        [http://dx.doi.org/10.1021/ja051340t] [PMID: 15926810]

[6]   Simsek, T. Pure YbB_6 nanocrystals: First time synthesis *via* mechanochemical method. *Adv. Powder Technol.,* **2019**, *30*(6), 1219-1225.
        [http://dx.doi.org/10.1016/j.apt.2019.03.018]

[7]   Bao, L-H.; Zhang, J-X.; Zhou, S-L.; Zhang, N.; Xu, H. Floating zone growth and thermionic emission property of single crystal CeB_6. *Chin. Phys. Lett.,* **2011**, *28*(8)088101
        [http://dx.doi.org/10.1088/0256-307X/28/8/088101]

[8]   Zheng, M.J.; Zhang, L.D.; Li, G.H.; Shen, W.Z. Fabrication and optical properties of large-scale uniform zinc oxide nanowire arrays by one-step electrochemical deposition technique. *Chem. Phys. Lett.,* **2002**, *363*(1-2), 123-128.
        [http://dx.doi.org/10.1016/S0009-2614(02)01106-5]

[9]   Engel, Y.; Elnathan, R.; Pevzner, A.; Davidi, G.; Flaxer, E.; Patolsky, F. Supersensitive detection of explosives by silicon nanowire arrays. *Angew. Chem. Int. Ed.,* **2010**, *49*(38), 6830-6835.
[http://dx.doi.org/10.1002/anie.201000847] [PMID: 20715224]

[10]   Zou, C.Y.; Zhao, Y.M.; Xu, J.Q. Synthesis of single-crystalline CeB_6 nanowires. *J. Cryst. Growth,* **2006**, *291*(1), 112-116.
[http://dx.doi.org/10.1016/j.jcrysgro.2006.02.042]

[11]   Ding, Q.; Zhao, Y.; Xu, J.; Zou, C. Large-scale synthesis of neodymium hexaboride nanowires by self-catalyst. *Solid State Commun.,* **2007**, *141*(2), 53-56.
[http://dx.doi.org/10.1016/j.ssc.2006.10.001]

[12]   Xu, J.; Zhao, Y.; Zou, C. Self-catalyst growth of LaB_6 nanowires and nanotubes. *Chem. Phys. Lett.,* **2006**, *423*(1-3), 138-142.
[http://dx.doi.org/10.1016/j.cplett.2006.03.049]

[13]   Zhang, H.; Zhang, Q.; Zhao, G.; Tang, J.; Zhou, O.; Qin, L.C. Single-crystalline GdB_6 nanowire field emitters. *J. Am. Chem. Soc.,* **2005**, *127*(38), 13120-13121.
[http://dx.doi.org/10.1021/ja054251p] [PMID: 16173720]

[14]   Xu, J.Q.; Zhao, Y.M.; Ji, X.H.; Zhang, Q.; Lau, S.P. Growth of single-crystalline SmB_6 nanowires and their temperature-dependent electron field emission. *J. Phys. D Appl. Phys.,* **2009**, *42*(13)135403
[http://dx.doi.org/10.1088/0022-3727/42/13/135403]

[15]   Xu, J.; Chen, X.; Zhao, Y.; Zou, C.; Ding, Q.; Jian, J. Self-catalyst growth of EuB_6 nanowires and nanotubes. *J. Cryst. Growth,* **2007**, *303*(2), 466-471.
[http://dx.doi.org/10.1016/j.jcrysgro.2006.12.040]

[16]   Bukatova, G.A.; Kuznetsov, S.A. Electrosynthesis of gadolinium hexaboride nanotubes. *Electrochem. Commun.,* **2005**, *7*(6), 637-641.
[http://dx.doi.org/10.1016/j.elecom.2005.04.003]

[17]   Yi, G.C.; Wang, C.; Park, W.I. ZnO nanorods: synthesis, characterization and applications. *Semicond. Sci. Technol.,* **2005**, *20*(4), S22-S34.
[http://dx.doi.org/10.1088/0268-1242/20/4/003]

[18]   Jha, M.; Patra, R.; Ghosh, S.; Ganguli, A.K. Vertically aligned nanorods of lanthanum hexaboride with efficient field emission properties. *Solid State Commun.,* **2013**, *153*(1), 35-39.
[http://dx.doi.org/10.1016/j.ssc.2012.10.007]

[19]   Aprea, A.; Maspero, A.; Masciocchi, N.; Guagliardi, A.; Albisetti, A.F.; Giunchi, G. Nanosized rare-earth hexaborides: Low-temperature preparation and microstructural analysis. *Solid State Sci.,* **2013**, *21*, 32-36.
[http://dx.doi.org/10.1016/j.solidstatesciences.2013.04.001]

[20]   Yu, Y.; Wang, S.; Li, W.; Chen, H.; Chen, Z. Synthesis of single-crystalline lanthanum hexaboride nanocubes by a low temperature molten salt method. *Mater. Chem. Phys.,* **2018**, *207*, 325-329.
[http://dx.doi.org/10.1016/j.matchemphys.2017.12.081]

[21]   Zhang, M.; Yuan, L.; Wang, X.; Fan, H.; Wang, X.; Wu, X.; Wang, H.; Qian, Y. A low-temperature route for the synthesis of nanocrystalline LaB_6. *J. Solid State Chem.,* **2008**, *181*(2), 294-297.
[http://dx.doi.org/10.1016/j.jssc.2007.12.011]

[22]   Kushkhov, H.B.; Vindizheva, M.K.; Mukozheva, R.A.; Abazova, A.H.; Tlenkopachev, M.R. Electrochemical synthesis of CeB_6 nanotubes. *J. Mater. Sci. Chem. Eng.,* **2014**, *2*(01), 57.

[23]   Kim, J.H.; Ko, Y.H.; Gong, S.H.; Ko, S.M.; Cho, Y.H. Ultrafast single photon emitting quantum photonic structures based on a nano-obelisk. *Sci. Rep.,* **2013**, *3*(1), 2150.
[http://dx.doi.org/10.1038/srep02150] [PMID: 23828558]

[24]   Brewer, J.R.; Deo, N.; Morris Wang, Y.; Cheung, C.L. Lanthanum hexaboride nanoobelisks. *Chem. Mater.,* **2007**, *19*(26), 6379-6381.

[http://dx.doi.org/10.1021/cm702315x]

[25]   Wang, G.; Brewer, J.R.; Chan, J.Y.; Diercks, D.R.; Cheung, C.L. Morphological evolution of neodymium boride nanostructure growth by chemical vapor deposition. *J. Phys. Chem. C,* **2009**, *113*(24), 10446-10451.
        [http://dx.doi.org/10.1021/jp901717h]

[26]   Mattox, T.; Urban, J. Tuning the Surface Plasmon Resonance of Lanthanum Hexaboride to Absorb Solar Heat: A Review. *Materials (Basel),* **2018**, *11*(12), 2473.
        [http://dx.doi.org/10.3390/ma11122473] [PMID: 30563148]

[27]   Chao, L-M.; Bao, L-H.; O, T. Optical response of CeB_6 nanoparticles with different sizes and shapes from discrete-dipole approximation. *Chin. Phys. Lett.,* **2015**, *32*(4)043301
        [http://dx.doi.org/10.1088/0256-307X/32/4/043301]

[28]   Simsek, T.; Avar, B.; Ozcan, S.; Kalkan, B. Nano-sized neodymium hexaboride: Room temperature mechanochemical synthesis. *Physica B,* **2019**, *570*, 217-223.
        [http://dx.doi.org/10.1016/j.physb.2019.06.047]

[29]   Bao, L.; Chao, L.; Li, Y.; Ming, M.; Yibole, B.; Tegus, O. SmB_6 nanoparticles: Synthesis, valence states, and magnetic properties. *J. Alloys Compd.,* **2015**, *651*, 19-23.
        [http://dx.doi.org/10.1016/j.jallcom.2015.06.086]

[30]   Ding, J.; Cui, C.; Sun, Y.; Ding, J.; Zhao, L.; Cui, S. Preparation of *in-situ* NdB_6 nanoparticles and their reinforcement effect on Al-Cu-Mn alloy. *J. Alloys Compd.,* **2019**, *806*, 393-400.
        [http://dx.doi.org/10.1016/j.jallcom.2019.07.237]

[31]   Schulz, M.J.; Kelkar, A.D.; Sundaresan, M.J. *Nanoengineering of structural, functional and smart materials*; CRC Press, **2005**.
        [http://dx.doi.org/10.1201/9780203491966]

[32]   Single Crystalline SmB_6 Nanostructure Arrays: Controllable Synthesis and Field Emission Property. *J. Inorg. Mater.,* **2020**, *35*(2), 199-204.

[33]   Gan, H.; Ye, B.; Zhang, T.; Xu, N.; He, H.; Deng, S.; Liu, F. A controllable solid-source CVD route to prepare topological Kondo insulator SmB_6 nanobelt and nanowire arrays with high activation energy. *Cryst. Growth Des.,* **2019**, *19*(2), 845-853.
        [http://dx.doi.org/10.1021/acs.cgd.8b01412]

[34]   Han, W.; Zhang, H.; Chen, J.; Zhao, Y.; Fan, Q.; Li, Q.; Liu, X.; Lin, X. Single-crystalline La Pr_1-B_6 nanoawls: Synthesis, characterization and growth mechanism. *Ceram. Int.,* **2016**, *42*(5), 6236-6243.
        [http://dx.doi.org/10.1016/j.ceramint.2016.01.006]

[35]   Zarzycki, J. *Glasses and the vitreous state*; , **1982**.

[36]   Wei, D.; Xie, J.; Tong, D.G. Amorphous Europium Hexaboride: A Potential Room Temperature Formaldehyde Sensing Material. *ACS Appl. Mater. Interfaces,* **2018**, *10*(42), 35681-35684.
        [http://dx.doi.org/10.1021/acsami.8b13234] [PMID: 30286288]

[37]   Ivashchenko, V.I.; Turchi, P.E.A.; Shevchenko, V.I.; Medukh, N.R.; Leszczynski, J.; Gorb, L. Electronic, thermodynamics and mechanical properties of LaB_6 from first-principles. *Physica B,* **2018**, *531*, 216-222.
        [http://dx.doi.org/10.1016/j.physb.2017.12.044]

[38]   Miller, J.C.; Serrato, R.; Represas-Cardenas, J.M.; Kundahl, G. *The handbook of nanotechnology: Business, policy, and intellectual property law*; John Wiley & Sons, **2004**.

[39]   Chen, B.; Yang, L.; Heng, H.; Chen, J.; Zhang, L.; Xu, L.; Qian, Y.; Yang, J. Additive-assisted synthesis of boride, carbide, and nitride micro/nanocrystals. *J. Solid State Chem.,* **2012**, *194*, 219-224.
        [http://dx.doi.org/10.1016/j.jssc.2012.05.032]

[40]   Ağaoğulları, D.; Balcı, Ö.; Öveçoğlu, M.L.; Suryanarayana, C.; Duman, İ. Synthesis of bulk nanocrystalline samarium hexaboride. *J. Eur. Ceram. Soc.,* **2015**, *35*(15), 4121-4136.

[http://dx.doi.org/10.1016/j.jeurceramsoc.2015.07.037]

[41] Wei, W.; Lihong, B.; Yingjie, L.; Luomeng, C.; Tegus, O. Solid-state reaction synthesis and characterization of PrB_6 nanocrystals. *J. Cryst. Growth,* **2015**, *415*, 123-126.
[http://dx.doi.org/10.1016/j.jcrysgro.2014.12.030]

[42] Wang, X.J.; Tian, J.F.; Yang, T.Z.; Bao, L.H.; Hui, C.; Liu, F.; Shen, C.M.; Xu, N.S.; Gao, H-J. Single crystalline boron nanocones: Electric transport and field emission properties. *Adv. Mater.,* **2007**, *19*(24), 4480-4485.
[http://dx.doi.org/10.1002/adma.200701336]

[43] Yang, X. Fabrication of single crystalline SmB_6 nanocone arrays and investigation of their field emission properties. *2016 29th International Vacuum Nanoelectronics Conference (IVNC),* **2016**, pp. 1-2.
[http://dx.doi.org/10.1109/IVNC.2016.7551503]

CHAPTER 4

The Rare-Earth Hexaborides Production Methods

Abstract: To produce rare-earth hexaborides, some methods exist: direct solid phase, carbothermal reduction, borothermal reduction, self-propagating synthesis, aluminum flux method, spark plasma sintering, and mechanochemical synthesis, floating zone method, and chemical vapor deposition. In this section, the drawbacks and advantages of these production methods will be discussed.

Keywords: Production methods, raw materials, advanced materials.

4.1. INTRODUCTION

To prepare pure REB_6 powders, various methods, such as direct solid phase, carbothermal reduction, borothermal reduction, self-propagating synthesis (SHS), aluminum flux method, spark plasma sintering (SPS), mechanochemical synthesis, floating zone method (FZM), and chemical vapor deposition (CVD), were used [1 - 4]. The single-phased REB_6 is studied by the direct solid-phase technique of $NaBH_4$ with CeO_2 and Eu_2O_3. This technique indicates the benefits of inexpensive and controlled grain dimensions. Moreover, this method is considered significant for the development of new RE nanomaterials with a wide variety of probable applications. The microstructure of raw materials needs to be examined because the particle size of raw materials in solid-state reactions affects the particle size of final products [5]. Currently, the carbothermal reduction technique is broadly improved because of its simple equipment and inexpensive. In addition, powders made with those techniques have a few defects like poorer sintering properties, lower naivety, and larger particle size [6]. In other studies, the average particle dimension of REB_6 produced by SHS by reduction technique shows that the grain size prepared by conventional technique is less than 500 nm which is finer. The SHS method is capable of producing materials with ultrafine microstructures, but grain growth occurs during the combustion reaction method because of high synthesis temperatures and improved mass transfer. It is likely to check the grain development by varying the parameters. In recent years, SHS has attracted attention because of its proven advantages, such as high product purity, low energy expenditure, and simple operation. Therefore, SHS method can be a good choice for producing REB_6 powders [7]. The aluminum flux method is also very important. This method has a significant function in the production of REB_6. But the fabrication productivity is small, and it is tough to abstain from the existence of Al contaminations [8]. SPS is a valuable technique for quick sintering, which can enhance the intensity and mechanical features. In addition, SPS is an important technique for the rapid intensification of ceramic nanopowders at low temperatures. The SPS method showed a proper technique to fabricate nanostructured REB_6 with superior properties [9]. Mechanochemical synthesis enables quick preparation for amorphous materials; oxide dispersion strengthened alloys, non-equilibrium alloys, advanced materials, nanocomposites, solid solution alloys, ceramics, and intermetallics, which are difficult or not possible to be acquired by traditional fabrication methods. Mechanochemical production is important in many characteristics, such as the type of milling atmosphere, milling time, milling rate, milling container, milling speed, milling environment, and ball-to-powder weight ratio (BPR) process control agent size, and dispersion of milling media. Mechanochemical synthesis is associated with fracturing, repeated welding, and the contact points between powder particles, providing favorable conditions for the presence of crops [10]. FZM is well suitable for preparing big refractory crystals. The development of big crystals of substances is possible with the floating zone technique. It is possible to examine the work function and the crystal electronic structure of REB_6 *via* FZM [11]. CVD method successfully produces REB_6 with well-defined morphology. REB_6 produced by the CVD is potentially used as dot electron emitters for field-based emission applications. The TEM, SEM, and heat field-based emission of electrons for smooth panel displays, like another electronic tools requires high-performance electron sources [12 - 15]. The self-propagating high-temperature synthesis method uses less energy for the fabrication of materials. Physical vapor deposition method (PVD) is corrosion resistant and high temperature resistant. SPS has many advantages, such as high sintering speed, high repeatability, and safety. The mechanochemical method is notable for its higher yields and shorter reaction times. The fabrication methods are summarized in Table **4.1** [16 - 33].

In this part, we have focused on the main methods that produce REB_6 structures. Basic information, recent trends, and new developments have been discussed.

4.2. PRODUCTION METHODS

4.2.1. Carbotermic Reduction Method

Carbotermic reduction is the phenomenon of reduction of metal-oxides to carbon and carbon intermediates. It is possible to examine the reactions in two groups direct reduction and indirect (indirect) reduction. Direct reduction events are reactions that occur as a result of the reduction of metal-oxides directly with carbon, while indirect reduction is the reaction that occurs as a result of the reduction of metal-oxides with carbon monoxide (CO), which is caused by the gasification of carbon (C). Raw and mechanically activated powder mixtures are carried out by the carbothermic reduction process in high-temperature furnaces according to the following stoichiometric ratios.

$$M_2O_3 + 3B_4C \rightarrow 2MB_6 + 3CO \text{ (M =La,Ca, Ce, Sm, Er and Eu)}$$

Table 4.1. The fabrication methods of the given materials.

Material	Fabrication Method	References
LaB_6	Carbotermic reduction method	[3]
PrB_6	Float zone method	[5]
CeB_6	Float zone method	[6]
CeB_6	Float zone method	[7]
LaB_6	Electrochemical Synthesis	[9]
CeB_6	Electrochemical Synthesis	[10]
LaB_6	Solid-State Reaction	[12]
PrB_6	Borothermal (Carbothermal)	[15]
CeB_6	Low-Temperature Synthesis	[19]
GdB_6	Low-Temperature Synthesis	[20]
LaB_6	Self-Propagating High Temperature	[22]
CeB_6	Self-Propagating High Temperature	[23]
LaB_6	Physical vapor deposition	[25]
LaB_6	Spark Plasma Sintering	[27]
LaB_6	Spark Plasma Sintering	[28]
CeB_6	Spark Plasma Sintering	[29]
CeB_6	Spark Plasma Sintering	[30]
LaB_6	Mechanochemical Synthesis	[32]
LaB_6	Mechanochemical Synthesis	[33]

Low-cost boron-carbide powders can be advantageously produced with low-temperature carbothermic reaction processes [1]. Detailed characterization studies of the obtained powders are carried out using XRD, SEM / EDS, TEM, and DSC tools. The carbothermal reduction production method has some disadvantages. Due to kinetic limitations such as limited contact area between reactants and irregular carbon distribution, the reaction may take place at higher temperatures. However, due to these limitations and high temperatures, grain growth, irregular grain shape, and unreacted carbon may be observed as a result of this process. It is a form of production with high energy consumption since it takes time to process [2]. Yu *et al.* synthesized LaB_6 nanoparticles using the carbothermic reduction method. The high purity of the synthesized LaB_6 was shown, and no additional peaks were detected from the impurities. The XRD analysis of high-purity LaB_6 powders is given in Fig. **(4.1)** [3].

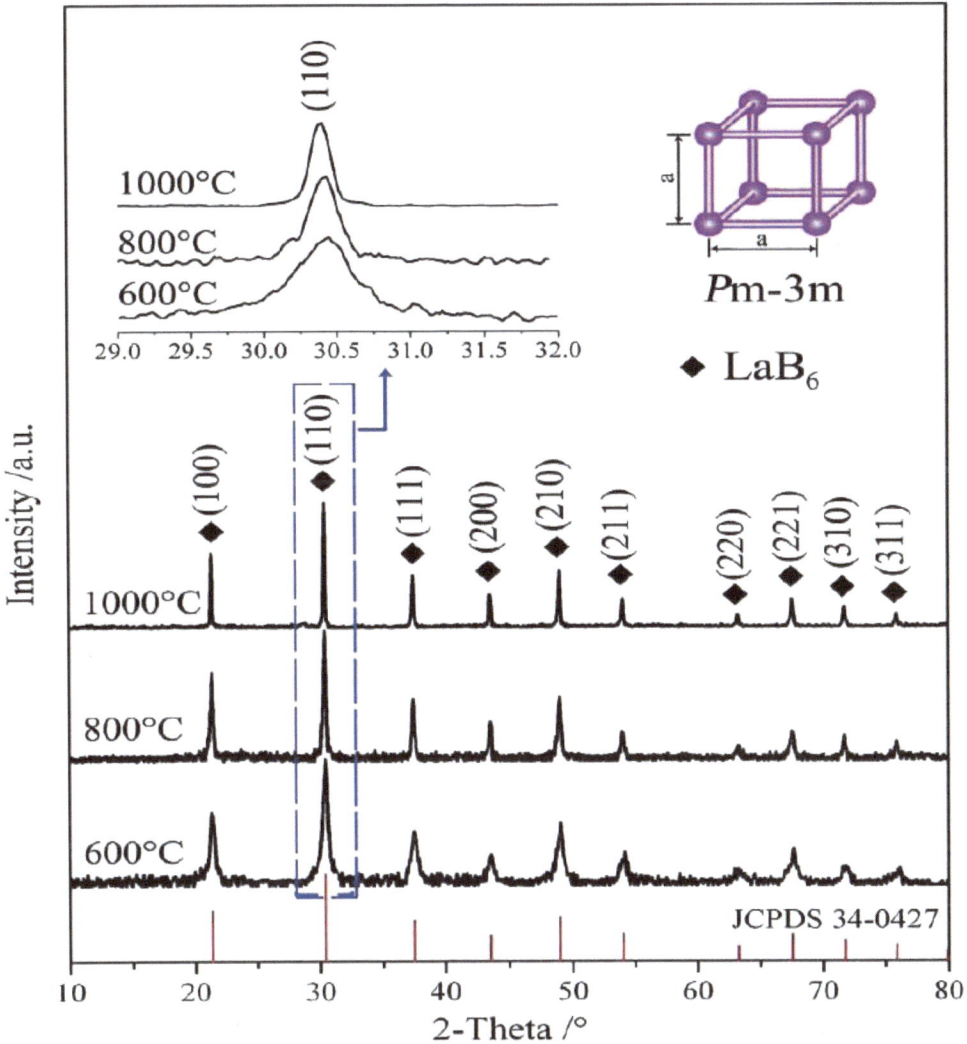

Fig. (4.1). XRD analysis of LaB$_6$ nanoparticles. Adopted from [3] (Copyright© 2017 Published by Elsevier B.V.).

4.2.2. Floating Zone Method (FZM)

The Floating zone method (FZM), invented by Theurer in 1962, is based on the principle of zone melting. The FZM production takes place under a vacuum or in an inert gas atmosphere. The process begins with a high-purity polycrystalline rod and a crystal of a monocrystalline seed that can be held face-to-face in a vertical position. With the help of a radio frequency, both are melted to some extent. The material (seed) is removed from the bottom up to make contact with the molten drop formed at the end of the poly rod. A neck stretching process is performed to form a dislocated crystal. When the molten region is moved along the polysilicon rod, the molten silicon solidifies into a single crystal, and simultaneously the material is purified [4]. Yu and coworkers produced the high-quality PrB_6 single crystal by an FZM. It was found that PrB_6 single crystals have a good application as a field emission cathode material. The XRD pattern of the single-crystal PrB_6 growth surface is shown in Fig. **(4.2)**. The diffraction peaks are indexed with the CsCl-type structure, and many sharped (100) peaks of the enlargement surface with no characteristic peaks of impurities are visibly shown, which affirms that the single crystal PrB_6 with (100) growth direction and a good purity have been produced by this technique. The growing PrB_6 single crystal was about 4 mm in diameter and 4 cm in length without pits, voids, cracks, or bubbles, existing on the crystal surface, as shown in Fig. **(4.2)** [5].

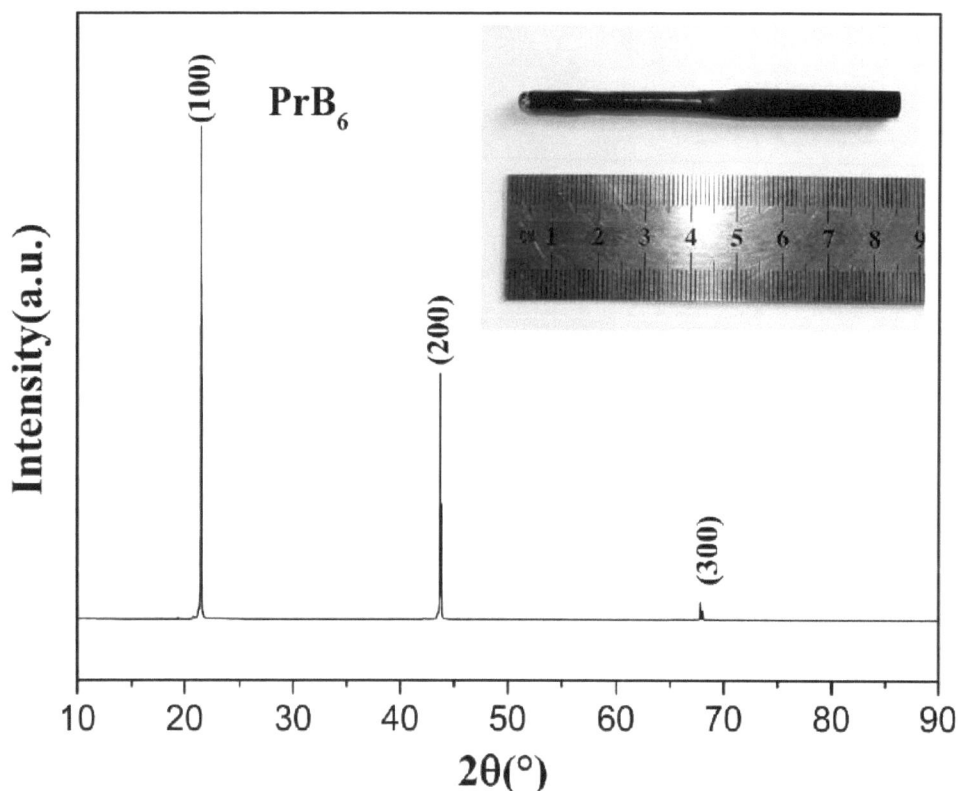

Fig. (4.2). XRD pattern of PrB_6 growth surface, insert photograph is the PrB_6 single crystal produced by FZM. Adopted from [5] (Copyright © 2018 Elsevier Ltd.).

In a different study by Hong *et al.*, successfully synthesized high-quality and large-sized CeB_6 single crystals were successfully grown by the optical FZM. The size of the produced CeB_6 single crystal is about 5 mm in diameter and 50 mm in length. The CeB_6 crystal growth was carried out in an optical floating zone furnace accoutred with four 3 kW Xenon lamps with a maximum heating temperature of 3000 °C. The furnace was perpendicularly configurated in such a method that the feed and seed rods were linearly aligned and can be rotated separately. The CeB_6 crystal growth was performed in an enclosed quartz tube, where a controlled Ar gas with a flow rate of 2 L/min and a pressure of 0.7 MPa was applied to decrease the vaporization rate during the growth. One advantage of the FZM is that it could supply a bigger temperature gradient on the growth interface, which is useful for stable growth [6]. In another study, Petrosyan successfully produced CeB_6 single crystals by FZM. Deviations from the presence and stoichiometry of uncontrolled impurities were shown in crystals grown by both the FZM and flux methods. Furthermore, they studied the interrelation between the chemical composition and thermoelectric properties of CeB_6 crystals synthesized by FZM and flux methods [7]. It is noteworthy that the use of this method is costly compared to other methods.

4.2.3. Electrochemical Synthesis

Electrochemical synthesis is one of the important techniques for the production of a variety of nanostructures and nanostructured energy substances, such as nanotubes, nanowires, nanorods, and composite nanostructures [8]. In a study, Wang *et al.* fabricated cold (field) emission sequences of sole crystal LaB_6 with uniform end structures employing the electrochemical etching technique and evaluated the cold emission properties. The applicability of electrochemical adhesion for the production of a single crystal LaB_6 field emitter array (FEA) is demonstrated by the SEM observation of emitters. The range between the LaB_6 surface and the graphite electrode was about 40 mm. The anode was implemented at ~3.0 V for 1.5 h and then reduced to ~1.5 V for 2 h to acquire sharp emitters. Single crystal LaB_6 FEA can be obtained effectively by applying good etching parameters [9]. Kushkhov *et al.* produced CeB_6 by electrochemical synthesis, as shown in Fig. (**4.3**). The results of this study showed that under specific circumstances, the concentrations of Ce and B and specific anionic compounds of the melt are likely for their joint electro reduction [10]. The efficiency of the single-phase CeB_6 was 0.20-0.30 g/A × hour. Certain surface field of ultra-dispersive powders of CeB_6 was 5-10 m^2/g.

Fig. (4.3). The SEM of CeB$_6$. Adoption from [10] (Copyright © 2006-2021 Scientific Research Publishing Inc.).

4.2.4. Solid-State Reaction

Solid-state reactions include reactions in which a change occurs in the atomic order, composition, or phase of the solid phase. For the most part, the reactions are exothermic. The interaction of solids depends on the structural defects of the solids. Solid-state reactions occur between powdered reagents. Molecules of solid substances do not move. If one or both crystals are heated, the gas phase can be released to form the product due to decomposition. The vapor pressure, density, particle size distribution, and porosity of the reagents play a very important role in solid-state reactions [11]. Zhang *et al.* successfully synthesized nanocrystalline LaB$_6$ using NaBH$_4$, LaCl$_3$, and metallic Mg powder at an autoclave temperature of 400 °C. The average particle of LaB$_6$ was 30 nm. They obtained LaB$_6$ nanotubes with an average size of ~200 nm at 500 °C using B$_2$O$_3$ instead of NaBH$_4$.

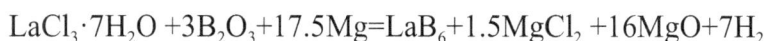

$$LaCl_3 \cdot 7H_2O + 6NaBH_4 + 4Mg = LaB_6 + 3NaCl + 4MgO + 3NaOH + 17.5H_2$$

$$LaCl_3 \cdot 7H_2O + 3B_2O_3 + 17.5Mg = LaB_6 + 1.5MgCl_2 + 16MgO + 7H_2$$

The FESEM picture of the LaB$_6$ nanocubes is shown in Fig. **(4.4)**. LaB$_6$ crystallines have an average particle size of ~200 nm [12].

Fig. (4.4). FESEM picture of the LaB_6 nanocubes. Adopted from [12] (Copyright© 2008 Elsevier Inc.).

The disadvantage is particle size, resulting from a solid-state synthesis reaction that has high-temperature limits for the use of the material.

4.2.5. Borothermal (Carbothermal) and Metallothermic (Aluminothermic) Reduction

Borothermal (Carbothermal) is the reduction that requires a very high temperature. Borothermic reduction is a production method used in the production of non-oxide boride, carbide, and nitrides, and can be applied by various methods. For the synthesis of borides, it is necessary to use a boron-containing carbon source or elemental boron as a boron source, in addition to carbon and metal oxide. In all cases, the borothermic reduction takes place as a result of endothermic reactions that require a lot of energy. CO gas occurs as a by-product. Borothermic reduction is possible at very high temperatures [13].

In aluminothermic (metallothermic) reactions, which are exothermal chemical reactions, iron (III) oxide (Fe_2O_3) is used. Also, high-temperature reduction agents such as Al are used. The method is industrially helpful for the fabrication of alloys of iron. The most major instance is the thermic reaction between aluminum and iron oxides to fabricate iron itself:

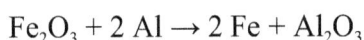

$$Fe_2O_3 + 2\ Al \rightarrow 2\ Fe + Al_2O_3$$

This definite reaction is, however, not related to the most significant application of aluminothermic reactions and the production of ferroalloys. For the fabrication of iron, a cheaper decreasing agent, coke, is performed by the carbothermic reaction. The aluminothermic reaction is performed for the production of some ferroalloys, for instance, ferroniobium from niobium pentoxide and ferrovanadium from aluminum, vanadium (V) oxide, and iron [14]. A study by Bliznakov *et al.*, successfully fabricated Ce and Pr by the reaction of Nd_2O_3, CeO_2, and Pr_6O_{11} with B, at temperatures between 1200 and 1800 °C in a vacuum according to the general scheme:

$$Me_mO_n + (6m+n)B \rightarrow mMeB_6 + nBO$$

It was found that reaction results of the compound close to stoichiometric are obtained in the range of 1700 and 1800°C. The RE (Nd, Ce, and Pr) and B_2O_3 are obtained by the borothermic reduction of oxides, the latter giving, with the excess of B, volatile B_6O. This reaction is an equilibrium process, relocated totally to BO at high temperatures. The crystal arrangements of PrB_6, CeB_6, and NdB_6 were defined with a good camera.

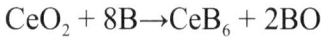

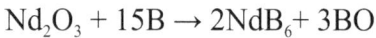

$$CeO_2 + 8B \rightarrow CeB_6 + 2BO$$

$$Pr_6O_{11} + 47B \rightarrow 6PrB_6 + 11BO$$

$$Nd_2O_3 + 15B \rightarrow 2NdB_6 + 3BO$$

Briefly, the implementation of the borothermic reduction technique employs the arrangement of good cleanliness products, also when general; "Moissan" or "amorphous" boron is employed to fabricate. When the initial supplies have been adequately pristine, the only contamination found in the last products can be MgB_2 [15]. Samsonov *et al.* successfully produced the development of REB_6 in the course of the reaction of metal oxides with B_4C, B, and a mixture of B and C in a vacuum.

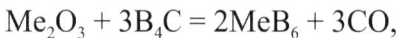

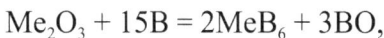

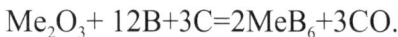

$$Me_2O_3 + 3B_4C = 2MeB_6 + 3CO,$$

$$Me_2O_3 + 15B = 2MeB_6 + 3BO,$$

$$Me_2O_3 + 12B + 3C = 2MeB_6 + 3CO.$$

Furthermore, in the vacuum high-temperature action, many of the impurities present in the initial boron and oxides are cleaned, and the final compounds are preserved against nitriding and oxidation. Their experimental results showed that the recommended RE production techniques yield adequately carbon-free products. Thus, these techniques are more advantageous than the earlier techniques, in which subtraction of carbon-containing contaminations from the products is inevitable [16].

4.2.6. Low-Temperature Synthesis in Autoclave or Reactor

Autoclaves are pressure-resistant boilers that can be adjusted to certain pressure and temperature values, using pressurized water steam for sterilization [17]. The parts in each autoclave are a) The boiler is the section where the sample is to be sterilized, b) Cover: the autoclave covers must be tightly closed with a screw, c) The thermostat ensures that the temperature is kept at a certain level, d) Air discharge cock: When the autoclave starts to operate, e) Pressure regulating valve: It is to keep the pressure constant by allowing the steam to escape after reaching a certain pressure level, f) Thermometer and manometer: The thermometer measuring the temperature and the manometer measuring the pressure show the values reached by the autoclave during its operation [18]. Wang *et al.* synthesized CeB_6 nanomaterials at low temperatures. Wang applied the TEM method to examine the elaborated morphology and crystal structure of CeB_6 nanowires (NWs) [19]. Fig. (4.5) displays a classic low enlargement TEM picture of produced CeB_6 NWs. The NWs shown with two arrows were seen as flat and smooth with a homogeneous diameter of 37 nm throughout the whole nanowires.

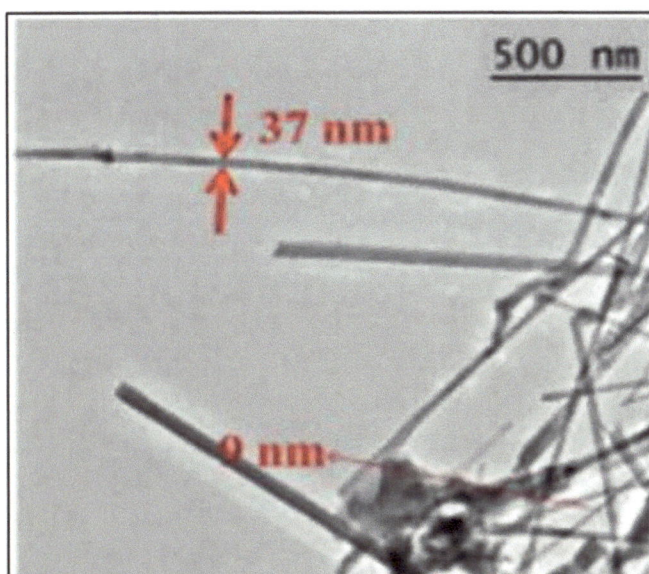

Fig. (4.5). TEM image of CeB_6 NWs. Adopted from [19] (Copyright © 2019 The Society of Powder Technology Japan. Published by Elsevier B.V.).

Han *et al.* synthesized high-quality single-crystalline GdB_6 nanowires in the range of 200 and 240 °C by a high-pressure solid-state (HPSS) technique in an autoclave where H_3BO_3, Gd, I_2, and Mg were used as raw materials. The morphology, valence, crystal structure, optical absorption, and magnetic properties were studied using HRTEM, superconducting quantum interference device (SQUID) magnetometry, XPS, XRD, FESEM, and optical measurements. SAED patterns and HRTEM images show that the GdB_6 nanowires are sole crystals with a good enlargement direction along (001). The XPS spectrum indicates that the value of the Gd ion in GdB_6 is trivalued. The optical features show strong adsorption in the NIR and UV range and weak adsorption in the visible light range. High NIR adsorption and low work function can make GdB_6 nanowires possible solar radiation-conserving substances for NIR blocking or solar cells. The chemical reaction of the synthesis of GdB_6 NWs is as follows:

$$Gd(s) + 6H_3BO_3(s) + 10\ Mg(s) + I_2(s) = GdB_6(s) + 9MgO(s) + MgI_2(s) + 9H_2O(g)$$

The SEM of GdB_6 nanostructures obtained at a low temperature of 200 °C is shown in Fig. **(4.6)**. The GdB_6 nanowires have diameters from 70 to 110 nm. It is seen that 200 °C is the lowest temperature to produce GdB_6 nanowires [20].

Fig. (4.6). SEM images of GdB_6 nanowires produced at 200 °C. Adopted from [20] (Copyright© 2017 Elsevier Inc.).

4.2.7. Self-Propagating High-Temperature Synthesis Method

Self-propagating high-temperature synthesis (SHS) is a technique used for the production of both inorganic and organic compositions with exothermic combustion reactions in solids of dissimilar structures [21]. The working principle is that the production of the materials is maintained by self-made heat from the chemical reaction. The reactants change into the resultant during the burning that takes place after the reaction begins. It has the advantages of short production cycles, low energy consumption, simple processes, and low costs. Intermetallics and composite substances can be prepared by utilizing this technique. By using this method, rare-earth hexaborides can be synthesized. For example, Jiang *et al.* fabricated LaB_6 ceramic powder from La_2O_3-B_2O_3-Mg by SHS with a reduction process. The combustion reaction in the inert atmosphere (Ar) was performed at $1600°C$. The yields were washed with distilled water and dilute HCL, respectively, to eliminate MgO, $LaBO_3$, $Mg_3B_2O_6$ and other contaminations. The effect of pressure and adding of SHS diluent on the grain dimension was studied. The result demonstrates that as the pressure enlarged, the average dimension of the powder reduced slowly. The addition of diluents decreased the produced temperature. Furthermore, the microstructure studies show that the average LaB_6 powder produced by SHS by reduction is less than 500 nm in grain size [22]. In another study, Zhihe *et al.* synthesized CeB_6 powders by which Mg, CeO_2 and B_2O_3 received qua reactants. The adiabatic temperature and dynamics of SHS reactions were studied. They characterized filtered products and SHS reaction products with XRD and SEM. Their results demonstrated that the adiabatic temperature of the Mg-B_2O_3-CeO_2 reaction system was quite bigger than 1800 K. The reaction orders (n) of Mg-B_2O_3-CeO_2 are 0.44, 1.31, and 1.36, respectively. Also, the apparent activation energies (Ea) of Mg-B_2O_3-CeO_2 are 14.88, 23.03, and 163.13 kJ/mol, respectively. The SHS yields are composed of CeB_6, MgO, and a small amount of $Mg_3B_2O_6$. The leached yields composed of the CeB_6 phase and its mean particle dimension of CeB_6 were lower than 150 nm. Also, its naivety was bigger than 99.0%. The XRD pattern of SHS products is shown in Fig. (**4.7**). SHS products are found to contain MgO and CeB_6 as well as a small amount of $Mg_3B_2O_6$. The reaction mechanism can be specified as:

$$CeO_2+3B_2O_3+11Mg=CeB_6+11MgO$$

$$3MgO+B_2O_3=Mg_3B_2O_6$$

CeB_6 is insoluble, but MgO and $Mg_3B_2O_6$ can be dissolved in HCL. Therefore the SHS products can be leached in a water bath at 50 °C with 6 mol/L HCl acid for more than 4 h with mechanical agitation of 200 r/min [23].

4.2.8. Physical Vapor Deposition (PVD)

Physical vapor deposition (PVD) methods are accumulation methods in which atoms or molecules of a substance are evaporated from a liquid or solid resource, moved in the form of vapor through low-pressure gaseous surroundings or a vacuum, and concentrated on a substratum. PVD processes could be used alloy, to deposit films of elemental, and compound materials as well as some polymeric materials [24]. Wu *et al.* fabricated LaB_6 films from DC magnetron sputtering technique. Magnetron spraying (sputtering) is a too significant PVD technique for using thin films due to the easiness of checking over the stoichiometry of the accumulated film. The fabrication of LaB_6 slim films *via* magnetron sputtering has received enormous attention because of the low substrate temperature and high deposition rates acquired in the sputtering route. Therefore, they have produced the LaB_6 target in five different types of doping to establish the structure and compound among the aim and film and produce slim films under similar spray process parameters to obtain the stoichiometric LaB_6 film for steady performance [25].

Fig. (4.7). The SEM images of the La_2O_3-B_2O_3-Mg powder. Adopted from [33] (Copyright © 2012 Elsevier Ltd and Techna Group S.r.l.).

4.2.9. Spark Plasma Sintering (SPS)

Bulk materials can be formed *via* sintering by employing different powder samples. SPS is an occurring powder consolidating method that ensures important benefits in the treating of high-temperature samples with weak deformation into configuring that was beforehand unapproachable. It is known as the pressure technique, which is activated by direct current, low-voltage, and pulsed current. It could be implemented to fabricate a wide variety of powders or bulk samples. In the sintering process, an electric field and exterior pressure are implemented at the same time to improve the intensity of the powder compacts. It can be applied to synthesized starting bulk samples from a wide variety of powders [26]. In another study conducted by Zhou *et al.*, nanostructured polycrystalline LaB_6 ceramics were produced with the SPS technique employing boron nanopowders and LaH_2 powders with particle dimensions of approximately 30 nm fabricated with H DC arc plasma. They examined the microstructure, crystal structure, grain orientations, and the reaction mechanism of sintering, and the features of the samples were studied using Neutron powder diffraction (NPD), Raman spectroscopy, XRD, differential scanning calorimetry (DSC), electron backscattered diffraction (EBSD) and TEM. The dense nanostructured LaB_6 with a fibrous texture is manufactured at a temperature of 1300 °C and with SPS at a pressure of 80 MPa for 5 minutes. The nanostructured LaB_6 bulk has both bigger thermionic emissions and bigger mechanical features. The flexural strength was 271.2 MPa, and the Vickers hardness was 22.3 GPa. In addition, the maximum emission current density at a cathode temperature of 1600 °C was 56.81 A cm^{-2} [27].

In another study conducted by Shenlin *et al.*, the high-density LaB_6 polycrystalline bulks were synthesized by SPS. The effects of pressure and sintering temperature on the microstructure and features of the sintering LaB_6 bulks were examined systematically. The sintering pressure of 50 MPa, sintering temperature of 1650 °C, and holding time of 10 mins are appropriate to the optimal technological parameters. They produced the emissive and mechanical features of the sintered LaB_6 bulks. The intensity of the LaB_6, Vickers hardness, the electron-emitting density, work function, and the highest bend strength is 96.2%, 1720 kg/mm^2, 17.41 A/cm^2, 2.40 eV, and 203.2 MPa, respectively. The SPS technique dampens the sintering time and reduces the sintering temperature of LaB_6 [28]. In a different study carried out by Lihong *et al.*, the polycrystalline CeB_6 cathode materials were synthesized by the SPS technique in the oxygen (O) system. They produced nanopowders with a mean grain size of 50 nm with high-energy ball milling. The ball-milled nanopowder was fully concentrated at 1550 °C under a pressure of 50 MPa. The Vickers hardness and flexural strengths were 51% and 211% higher than that of vulgar powder, respectively. The (100) fiber texture was produced at 1550 °C by the sintered ball-milled nanopowder, and the thermionic emission current density was measured as 16.04 A/cm² at 1873 K. In this study, the electrical resistance increases linearly from 36.34 to 60.54 μΩ•cm with a measurement temperature of 25 to 500 °C and exhibits a metallic conductive behavior. The electrical resistance of CeB_6 is higher than LaB_6. It is generally accepted that the ideal thermionic cathode sample is heated by self-resistance to electron emission, referred to as a "direct heating cathode". In addition, CeB_6 has a higher electrical resistance compared to LaB_6 and, therefore will be more suitable for a "direct heating cathode" [29]. Koroglu *et al.* fabricated polycrystal bulk CeB_6 ceramics by SPS employing CeO_2 and nano B dust. A two-stage heating program was implemented using FactSage thermochemical software, optimizing both synthesis and sintering temperatures. They studied the impact of boron particle dimension on the microstructural, physical, electrical, and mechanical properties of CeB_6 ceramics [30].

4.2.10. Mechanical Alloying (Mechanochemical Synthesis)

The mechanical alloying/milling (MA) route is a solid-state powder route where the powder particles are exposed to a highly energetic effect by the balls in a flask. As the powder particles in the flask are constantly affected by the balls, cold welding between particles and breaking of the particles come true repetitively during the ball milling route. The total process contains the blending of the powder mixture earlier to the ball milling, filling, or vacuuming with protective gases to preclude contamination and oxidation, and the ball milling route itself. Post-milling routes after mechanical alloying contain degassing, canning, and finally, plastic deformation by rolling or extrusion. For the milling route, several control parameters have to be considered, namely, the ball-to-powder weight ratio, the environment in the flask, and process control agents [31]. Tekoglu *et al.* fabricated LaB_6 particulate reinforced Al-12.6 wt.% Si composites by MA and spark plasma sintering. They examined the physical, microstructural, and mechanical properties of the structures. They performed MA for 2, 4, and 8 hours using a high-energy ball mill (with a speed of 1060 rpm). The highest hardness $(1.72 \pm 0.12$ GPa) and yield strength (373 MPa) were obtained in the structure of 10 wt % LaB_6-Al-12.6 wt % Si composites [32]. In another study, Agaogulları *et al.* produced LaB_6 powders from mixtures of La_2O_3, B_2O_3, and Mg in a mechanochemical method [33]. Fig. **(4.7)** exhibit the SEM images of the La_2O_3-B_2O_3-Mg. As seen in Fig. **(4.7)**, as-blended powders have white irregular agglomerates and dark gray leaf-like particles. Blended powders have white erratic agglomerates and dark gray leaves as particles.

CONCLUSION

REB$_6$ materials can be produced by various methods such as direct solid phase, carbothermal reduction, borothermal reduction, self-propagating synthesis, aluminum flux method, spark plasma sintering, mechanochemical synthesis, floating zone method, and chemical vapor deposition. The single-phased REB$_6$ was examined by the direct solid-phase technique, which shows the benefits of inexpensive and controlled grain dimensions. Furthermore, this method is used for the development of new RE nanomaterials. Currently, the carbothermal reduction technique can be preferred due to its simple equipment and inexpensive. The SHS method is capable of synthesizing ultrafine microstructured materials, but grain growth occurs during the combustion reaction method. SPS is valuable for rapid sintering. Mechanochemical synthesis enables quick preparation for amorphous materials; oxide dispersion strengthened alloys, non-equilibrium alloys, advanced materials, nanocomposites, solid solution alloys, ceramics, and intermetallics FZM is well suitable for preparing big refractory crystals. The development of big crystals of substances is possible with the floating zone technique. The CVD method successfully produces REB$_6$ with well-defined morphology. REB$_6$ produced by the CVD is potentially used as dot electron emitters for field-based emission applications. The physical vapor deposition method (PVD) is corrosion-resistant and high-temperature resistant. SPS has many advantages, such as high sintering speed, high repeatability, and safety.

REFERENCES

[1]   Gasch, M.J.; Ellerby, D.T.; Johnson, S.M. Ultra high-temperature ceramic composites. In: *Handbook of ceramic composites*; Springer, **2005**; pp. 197-224.
[http://dx.doi.org/10.1007/0-387-23986-3_9]

[2]   Jiang, Y.; Li, R.; Zhang, Y.; Zhao, B.; Li, J.; Feng, Z. Tungsten doped ZrB$_2$ powder synthesized synergistically by co-precipitation and solid-state reaction methods. *Procedia Eng.,* **2011**, *27*, 1679-1685.
[http://dx.doi.org/10.1016/j.proeng.2011.12.636]

[3]   Yu, Y.; Wang, S.; Li, W.; Chen, Z. Low temperature synthesis of LaB$_6$ nanoparticles by a molten salt route. *Powder Technol.,* **2018**, *323*, 203-207.
[http://dx.doi.org/10.1016/j.powtec.2017.09.049]

[4]   Meroli, S. Two growth techniques for monocrystalline silicon: Czochralski *vs.* Float Zone. **2012**.

[5]   Liu, H.; Zhang, X.; Xiao, Y.; Wang, Y.; Zhang, J. The thermionic and field emission properties of single crystal PrB$_6$ grown by floating zone method. *Vacuum,* **2018**, *151*, 76-79.
[http://dx.doi.org/10.1016/j.vacuum.2018.02.015]

[6]   Bao, L-H.; Zhang, J-X.; Zhou, S-L.; Zhang, N.; Xu, H. Floating zone growth and thermionic emission property of single crystal CeB$_6$. *Chin. Phys. Lett.,* **2011**, *28*(8), 088101.
[http://dx.doi.org/10.1088/0256-307X/28/8/088101]

[7]   Petrosyan, V.; Vardanyan, V.; Kuzanyan, V.; Konovalov, M.; Gurin, V.; Kuzanyan, A. Thermoelectric properties and chemical composition of CeB$_6$ crystals obtained by various methods. *Solid State Sci.,* **2012**, *14*(11-12), 1653-1655.
[http://dx.doi.org/10.1016/j.solidstatesciences.2012.05.030]

[8]   Li, G.R.; Xu, H.; Lu, X.F.; Feng, J.X.; Tong, Y.X.; Su, C.Y. Electrochemical synthesis of nanostructured materials for electrochemical energy conversion and storage. *Nanoscale,* **2013**, *5*(10), 4056-4069.
[http://dx.doi.org/10.1039/c3nr00607g] [PMID: 23584514]

[9]   Wang, X.; Jiang, Y.; Lin, Z.; Qi, K.; Wang, B. Field emission characteristics of single crystal LaB$_6$ field emitters fabricated by electrochemical etching method. *J. Phys. D Appl. Phys.,* **2009**, *42*(5), 055409.
[http://dx.doi.org/10.1088/0022-3727/42/5/055409]

[10]   Kushkhov, H.B.; Vindizheva, M.K.; Mukozheva, R.A.; Abazova, A.H.; Tlenkopachev, M.R. Electrochemical synthesis of CeB$_6$ nanotubes. *J. Mater. Sci. Chem. Eng.,* **2014**, *2*(01), 57.

[11]   Durak, D. *Synthesis and Characterization of Some Metal Borates*; Master Thesis, Kırıkkale University, Institute of Science and Technology, Kırıkkale, **2007**.

[12]   Zhang, M.; Yuan, L.; Wang, X.; Fan, H.; Wang, X.; Wu, X.; Wang, H.; Qian, Y. A low-temperature route for the synthesis of nanocrystalline LaB$_6$. *J. Solid State Chem.,* **2008**, *181*(2), 294-297.
[http://dx.doi.org/10.1016/j.jssc.2007.12.011]

[13]   Carbide, N. *Boride Materials-Synthesis and Processing*; Weimer, A.W., Ed.; Chapman& Hall: London, **1997**.

[14]   R. Fichte, "Ferroalloys," *Ullmann's Encycl. Ind. Chem.*, **2000**.
[http://dx.doi.org/10.1002/14356007.a10_305]

[15]   Bliznakov, G.; Peshev, P. The preparation of cerium, praseodymium, and neodymium hexaborides. *J. Less Common Met.,* **1964**, *7*(6), 441-446.
[http://dx.doi.org/10.1016/0022-5088(64)90041-4]

[16]   Samsonov, G.V.; Paderno, Y.B.; Fomenko, V.S. Hexaborides of the rare-earth metals. *Soviet Powder Metallurgy and Metal Ceramics,* **1964**, *2*(6), 449-454.
[http://dx.doi.org/10.1007/BF00774188]

[17]   Estridge, B.H.; Reynolds, A.P.; Walters, N.J. *Basic medical laboratory techniques*; Cengage Learning, **2000**.

[18]   Karadağ, A. *Sterilization by autoclave,* 4th edition; , **2005**.

[19]   Wang, Z.; Han, W.; Kuang, Q.; Fan, Q.; Zhao, Y. Low-temperature synthesis of CeB$_6$ nanowires and nanoparticles as feasible lithium-ion anode materials. *Adv. Powder Technol.,* **2020**, *31*(2), 595-603.
[http://dx.doi.org/10.1016/j.apt.2019.11.014]

[20]   Han, W.; Wang, Z.; Li, Q.; Liu, H.; Fan, Q.; Dong, Y.; Kuang, Q.; Zhao, Y. Autoclave growth, magnetic, and optical properties of GdB$_6$ nanowires. *J. Solid State Chem.,* **2017**, *256*, 53-59.
[http://dx.doi.org/10.1016/j.jssc.2017.08.026]

[21]   Subrahmanyam, J.; Vijayakumar, M. Self-propagating high-temperature synthesis. *J. Mater. Sci.,* **1992**, *27*(23), 6249-6273.
[http://dx.doi.org/10.1007/BF00576271]

[22]   Jiang, N.; Wang, W.M.; Fu, Z.Y.; Wang, H.; Wang, Y.C.; Zhang, J.Y. Influence of preparation parameter on the grain size of LaB$_6$ powder synthesized by SHS with reduction process. *Adv. Mat. Res.,* **2010**, *105*, 351-354.

[23]   Dou, Z.; Zhang, T.; Liu, Y.; Guo, Y.; He, J. Preparation of CeB$_6$ nano-powders by self-propagating high-temperature synthesis (SHS). *J. Rare-Earths,* **2011**, *29*(10), 986-990.
[http://dx.doi.org/10.1016/S1002-0721(10)60583-2]

[24]   Mattox, D.M. Physical vapor deposition (PVD) processes. *Met. Finish.,* **2002**, *100*, 394-408.
[http://dx.doi.org/10.1016/S0026-0576(02)82043-8]

[25]   Wu, Y.; Min, G.; Chen, D.; Zhang, L.; Yu, H. The correlation of stoichiometry between boron-rich

LaB_6 targets and LaB_6 films. *Ceram. Int.,* **2015**, *41*(1), 1005-1013.
[http://dx.doi.org/10.1016/j.ceramint.2014.09.021]

[26] Cavaliere, P.; Sadeghi, B.; Shabani, A. Spark plasma sintering: process fundamentals. In: *Spark Plasma Sintering of Materials*; Springer, **2019**; pp. 3-20.
[http://dx.doi.org/10.1007/978-3-030-05327-7_1]

[27] Zhou, S.; Zhang, J.; Liu, D.; Lin, Z.; Huang, Q.; Bao, L.; Ma, R.; Wei, Y. Synthesis and properties of nanostructured dense LaB_6 cathodes by arc plasma and reactive spark plasma sintering. *Acta Mater.,* **2010**, *58*(15), 4978-4985.
[http://dx.doi.org/10.1016/j.actamat.2010.05.031]

[28] Shenlin, Z.; Jiuxing, Z.; Danmin, L. Properties of high density LaB_6 cathode prepared by spark plasma solid phase sintering. *High Power Laser and Particle Beams,* **2010**, *22*(1), 171-175.
[http://dx.doi.org/10.3788/HPLPB20102201.0171]

[29] Bao, L.; Zhang, J.; Zhou, S. Effect of particle size on the polycrystalline CeB_6 cathode prepared by spark plasma sintering. *J. Rare-Earths,* **2011**, *29*(6), 580-584.
[http://dx.doi.org/10.1016/S1002-0721(10)60501-7]

[30] Koroglu, L.; Ayas, E. *In-situ* synthesis and densification of CeB_6 ceramics by spark plasma sintering from CeO_2 and B powders: Effect of boron content and boron particle size on microstructural, mechanical and electrical properties. *Mater. Chem. Phys.,* **2020**, *240*, 122253.
[http://dx.doi.org/10.1016/j.matchemphys.2019.122253]

[31] Lü, L.; Lai, M.O. *Mechanical alloying*; Springer Science & Business Media, **2013**.

[32] Tekoğlu, E.; Ağaoğulları, D.; Yürektürk, Y.; Bulut, B.; Lütfi Öveçoğlu, M. Characterization of LaB_6 particulate-reinforced eutectic Al-12.6 wt% Si composites fabricated *via* mechanical alloying and spark plasma sintering. *Powder Technol.,* **2018**, *340*, 473-483.
[http://dx.doi.org/10.1016/j.powtec.2018.09.055]

[33] Ağaoğulları, D.; Duman, İ.; Öveçoğlu, M.L. Synthesis of LaB_6 powders from La_2O_3, B_2O_3 and Mg blends *via* a mechanochemical route. *Ceram. Int.,* **2012**, *38*(8), 6203-6214.
[http://dx.doi.org/10.1016/j.ceramint.2012.04.073]

CHAPTER 5

The Rare-Earth Hexaboride-Based Alloys

Abstract: The rare-earth hexaboride can be both alloyed with alkaline earth hexaboride and rare-earth hexaborides. Both alloying types have different types of advantages. For example, large-size triple $La_xCe_{1-x}B_6$ single crystals produced by the floating zone method showed excellent field emission and thermionic emission characteristics. Thus, these types of alloys indicate superior performance (electronic, magnetic, excellent field emission, thermionic emission properties) when compared to their pure counterparts.

Keywords: Alkaline earth materials, optical performance, the density of state.

5.1. INTRODUCTION

The alloys are formed by mixing two or more elements [1]. Alloying elements are added to REB_6 to induce ductility, hardness, toughness, or other desired properties. Most alloys can be work-hardened by creating defects in their crystal structure. These defects are created during plastic deformation by hammering, bending, and extruding. The properties of many alloys can also be changed by heat treatment. Some of the metals can be softened by annealing, which recrystallizes the alloy and repairs the defects [2]. Alloying REB_6 with other metals improves the properties of REB_6. For example, the alloying reduces the total kinetic energy of the electrons of REB_6 and leads to the absorption valleys altering to a longer wavelength direction and the red-shift in the absorption valley [3]. The density of states (DOS) of a system defines the proportion of states that will be occupied by the system at each energy. Fermi energy is the difference in energy between the highest and lowest occupied single-particle states in a quantum system that is usually composed of fermions that don't interact at absolute zero temperature in quantum mechanics. The alloying of REB_6 can adjust DOS and the position of the Fermi energy level. In addition, the work function of REB_6 can be considerably reduced, providing better emission performance than REB_6. They also can considerably improve their thermionic emission property [4]. The optical properties of alloyed REB_6 draw attention. Plasmon energy is proportional to the square root of the free electron density in a metal. The plasma frequency is the frequency at which electrons in the plasma naturally oscillate

Mikail Aslan and Cengiz Bozada

relative to ions and has values between 2 and 20 MHz—excited as a collective oscillation of all valence electrons also in semiconductors and insulators. Among its optical properties, plasmon energy and plasma frequency behavior stand out. Alloying of REB_6 leads to a reduction of the plasmon energy and plasma frequency of REB_6. REB_6 has higher Fermi energy than alloyed REB_6 [5].

In this part, we have focused on the two types of REB_6-based alloy structures. The chemical and physical properties of these alloys have been discussed.

5.2. THE ALLOYED ALKALINE-EARTH METAL HEXABORIDES MB_6 (M=CA, SR, BA) WITH RARE-EARTH HEXABORIDES

The electrical and mechanical properties of both alkaline and rare-earth hexaborides are important. Alkaline earth metal hexaboride (MB_6), CsCl type simple cubic CaB_6, SrB_6, and BaB_6 have common properties with Rare-Earth hexaborides due to their high hardness, high melting points, good chemical stability, low thermal expansion coefficient, and low density. The mechanical properties of these hexaborides are very significant because of their use as structural ceramics [6].

Lanthanum hexaboride (LaB_6) is characterized by its high melting point, high electrical conductivity, and low operating function [7, 8], making it one of the best thermionic materials for high electron-density cathodes. LaB_6 nanoparticles are used as solar radiation heat protection material for automotive and architectural windows due to their high optical absorption coefficient in the near-infrared region and high permeability in the visible region [9]. In another study, Takeda *et al.* prepared LaB_6 nanoparticles by a ball milling method to obtain ultra-fine nanoparticles [10]. Lihong Bao *et al.* conducted a synthesis of $La_{1-x}Ba_xB_6$, a cubic-shaped triple nanocrystalline, by simple step solid-state reaction, and the effects of Ba-doping on the optical and magnetic properties were investigated. Interestingly, Ba doping caused the wavelength of the absorption valley and the absorption peak of LaB_6 to shift to red. In addition, Ba doping causes the fermentation of nanocrystalline LaB_6 at room temperature. According to the first principle calculation results, the doped Ba reduces the total kinetic energy of electrons of LaB_6, so the absorption valleys move towards a higher wavelength [11]. CeB_6 shows excellent optical absorption properties [12]. Xiaoping Qi *et al.*, nanocrystalline Ca-doped CeB_6 powders were synthesized under a vacuum condition with $NaBH_4$ in a solid-state reaction of CaO and CeO_2. The optical absorption properties and grain morphology of nanocrystalline CeB_6 and Ca doping effects on phase composition were investigated by Xiaoping Qi. Grain size and morphology are very sensitive to the reaction temperature. As a result of optical absorption, the absorption valley of the nanocrystalline CeB_6

shows a redshift from 619 nm to 685 nm with increasing Ca doping; this means adjustable optical absorption of nanocrystalline Ca doped CeB_6. In addition, a first principle calculation was used to reveal the origin of adjustable optical properties. Ca doping was found to reduce the total kinetic energy of the CeB_6 electrons and cause the absorption valleys to shift to the longer wavelength direction [13]. Fig. (**5.1a**) shows the XRD models of the nanocrystalline $Ce_{0.8}Ca_{0.2}B_6$ prepared at a reaction temperature of 900-1200°C. Fig. (**5.1b**) shows the XRD patterns of nanocrystalline $Ce_{1-x}Ca_xB_6$ with various Ca doping contents x = 0, 0.2, 0.4, 0.6 and 0.8 prepared at 1200 ° C for 2 hours.

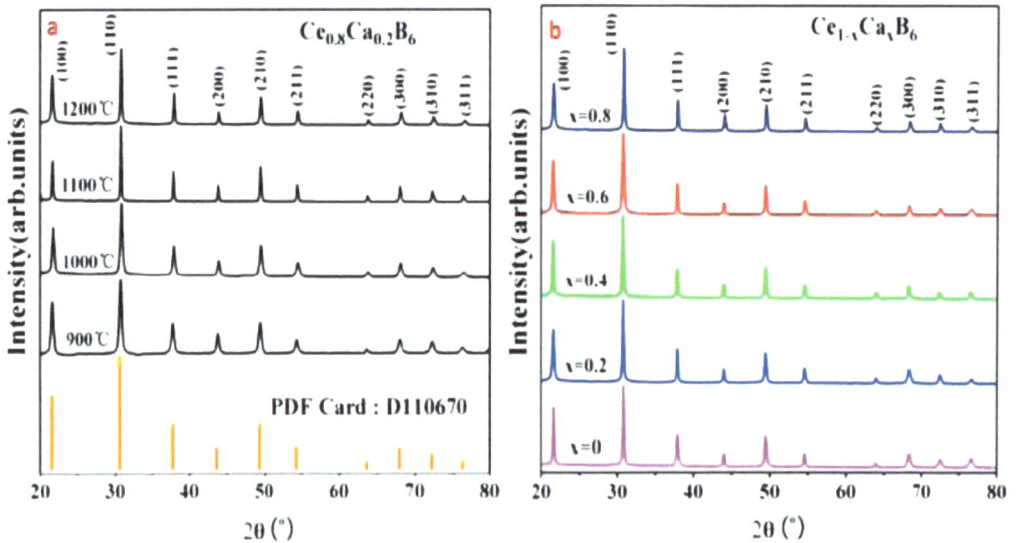

Fig. (**5.1**). (**a**) XRD patterns of nanocrystalline $Ce_{0.8}Ca_{0.2}B_6$ prepared at different temperatures for 2 hours; (**b**) XRD patterns of nanocrystalline $Ce_{1-x}Ca_xB_6$ prepared at 1200 ° C for 2 hours. Adopted from [13] (Copyright © 2017 Elsevier B.V.).

The origin of ferromagnetism in light electron-doped $Ca_{1-x}La_xB_6$ is a major problem in physics because there are no partially filled d or f orbitals required for magnetism [14]. In another study, EuB_6 has been extensively studied for its magnetic and transport properties for ferromagnetic transmission [15]. The ferromagnetism of EuB_6 was confirmed by neutron scattering measurements, and it was found that the magnetic moment stable Eu^2 moment was due to localized 4 f electrons [16]. In another study, Jong-Soo Rhyme *et al.* studied the temperature and field-dependent magnetic properties of the $Eu_{1-x}Ca_xB_6$ compounds. It was found that a ferromagnetic transition temperature up to Tc = 5.5 K is suppressed, and also, small Ca doping in $Eu_{0.87}Ca_{0.13}B_6$ caused the suppression of a ferromagnetic transition temperature from Tc = 12 K for EuB_6 to Tc = 5.5 K for $Eu_{0.87}Ca_{0.13}B_6$ [17]. Also, M. Batkova *et al.* conducted a study on the effect on the electrical resistivity of $EuB_{5.99}C_{0.01}$, which was considered internally non-homogeneous due to external pressure and fluctuations in carbon content. M. Batkova's research results show that low temperature resistant maximum shifts to a lower temperature with applied pressure, contrary to the reported behavior for stoichiometric EuB_6 [18].

5.3. THE ALLOYED RARE-EARTH HEXABORIDES

Rare-earth hexaborides (REB_6) have attracted great attention due to their electronic, magnetic, excellent field emission, and thermionic emission properties. The crystal structure, crystal quality, and emission characteristics of the alloyed REB_6 were characterized by diffractometer, X-ray Laue diffraction, electron paramagnetic resonance (EPR), and thermal electron emission measurements. The alloyed REB_6 demonstrates superior performance (electronic, magnetic, excellent field emission, thermionic emission properties) better than REB_6. Some of them are used as hot and cold cathodes. Also, the alloyed REB_6 show antiferromagnetic for x=3 and but when x<0.05, the alloyed REB_6 materials indicate ferromagnetic and metallic. This enables use in new optoelectronic applications.

Simple cubic CsCl-type rare-earth hexaborides have attracted great attention due to their electronic and magnetic properties [19]. In another study, Chao *et al.* investigated the optical performance of The Yb-contributed LaB_6 through the first principle calculation. They found that LaB_6, which coincided with the change of permeability peak from LaB_6 to Yb-contributed LaB_6, lowered its plasmon energy. This can be attributed to the reduction of Fermi energy. When some of the La atoms in LaB_6 are replaced by Yb atoms, the total kinetic energy of electrons is reduced; the number of load carriers and plasma frequency is also reduced, resulting in the minimum absorption value shifting towards the long wave direction. This adjustable feature of the permeability summit can expand the optical applications of rare soil hexaborides as optical filters [20]. In another study conducted by Mattox *et al.* Yb doped LaB_6 was investigated. Matrox used the first fundamental calculations of density functional theory (DFT) to study the potential optical effects of LaB_6 nanoparticles doped with ytterbium (Yb) and particles with ytterbium (Yb) and found that the 4f states of the Yb dopant on the Fermi surface had the potential to uniquely affect optical features due to the involvement of f-orbitals [21]. In the energy loss spectrum, a comparison of pure YbB_6, LaB_6, and $La_{0.625}Yb_{0.375}B_6$ phases is shown in Fig. (**5.2**). Doping LaB_6 with Yb reduces the plasmon energy of the system. The concentration of charge carriers in the system reduces and decreases the plasma frequency when a few of the La atoms are replaced with Yb, which reasoning the minimum plasma absorption to shift to higher wavelengths.

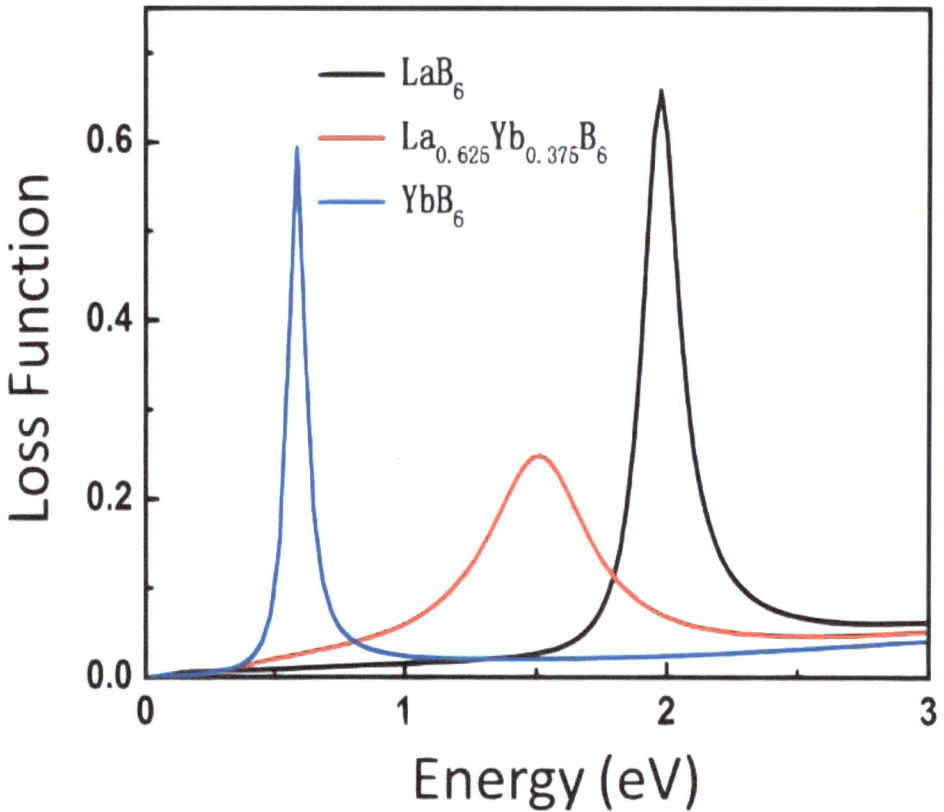

Fig. (5.2). The energy loss function of YbB_6, LaB_6, and $La_{0.625}Yb_{0.375}B_6$ in the low energy region. Adopted from [21] (Copyright © 2018 MDPI, Basel, Switzerland).

In another study, Bao *et al.* experimentally studied the optical absorption and structure of ternary nanocrystalline Eu doped SmB_6 powders (Fig. **5.3**). According to their optical absorption results, the transmittance wavelength of nanocrystalline SmB_6 increases from 696 nm to 1000 nm, while the Eu doping content increased to x=0.8. The transmittance wavelength adjustable mechanisms of nanocrystalline Eu-doped SmB_6, the density of states, the electronic structure, and the energy loss function were calculated by first-principle calculation. Eu^{2+} doping into SmB_6 caused the plasma frequency to effectively reduce the excitation energy from 1.48 to 1.25 eV and led to the transmittance wavelength redshift toward a higher wavelength [22].

Fig. (5.3). FESEM images of nanocrystalline $Sm_{1-x}Eu_xB_6$ powders prepared at 1150 °C at x = 0.8. Adopted from [22] (Copyright © 2020 Chinese Society of rare-earths. Published by Elsevier B.V.).

In a different study, Yeo *et al.* investigated the Eu-doping effects in SmB_6 based on measurements of magnetic susceptibility, resistance, and hall effect. SmB_6 is a Kondo insulator with a small gap caused by hybridization between broad conduction bands and a narrow f band. EuB_6 is a polaronic ferromagnet. $Sm_{1-x}Eu_xB_6$ crystals have a single-phase cubic structure with the space group Pm3m.

Along the cubic axis in a commercial superconducting quantum-interference device magnetometer, the magnetic properties of $Sm_{1-x}Eu_xB_6$ were measured by Yeo since Eu reduces the strength of hybridization between a narrow f band and broad conductive bands; they have created an AF supercharge interaction between magnetically active ions, which leads to the transition from the Kondo insulator to an AF insulator at x ≤ 0.2. Fig. (**5.4a**) describes the hunchback of ~55 K in the activation gap data. In Fig. (**5.4b**), it is seen that Tχ decreases more than 2 times when the temperature is reduced from 60 to 2K [23]. SmB_6 is a well-known Kondo insulator, where hybridization of mobile 5d electrons with localized 4f electrons leads to the transition from metallic to isolation behavior at low temperatures. Recent research shows that SmB_6 is a topological metallic surface topological insulator resulting from a completely insulating hybridized bulk tape structure. Akintola *et al.* investigated the magnetic properties of SmB_6 with 0,5% Fe added by muon rotation/relaxation methods. Fe impurity below 6 K triggers simultaneous changes in mass local magnetism and electrical conductivity [24].

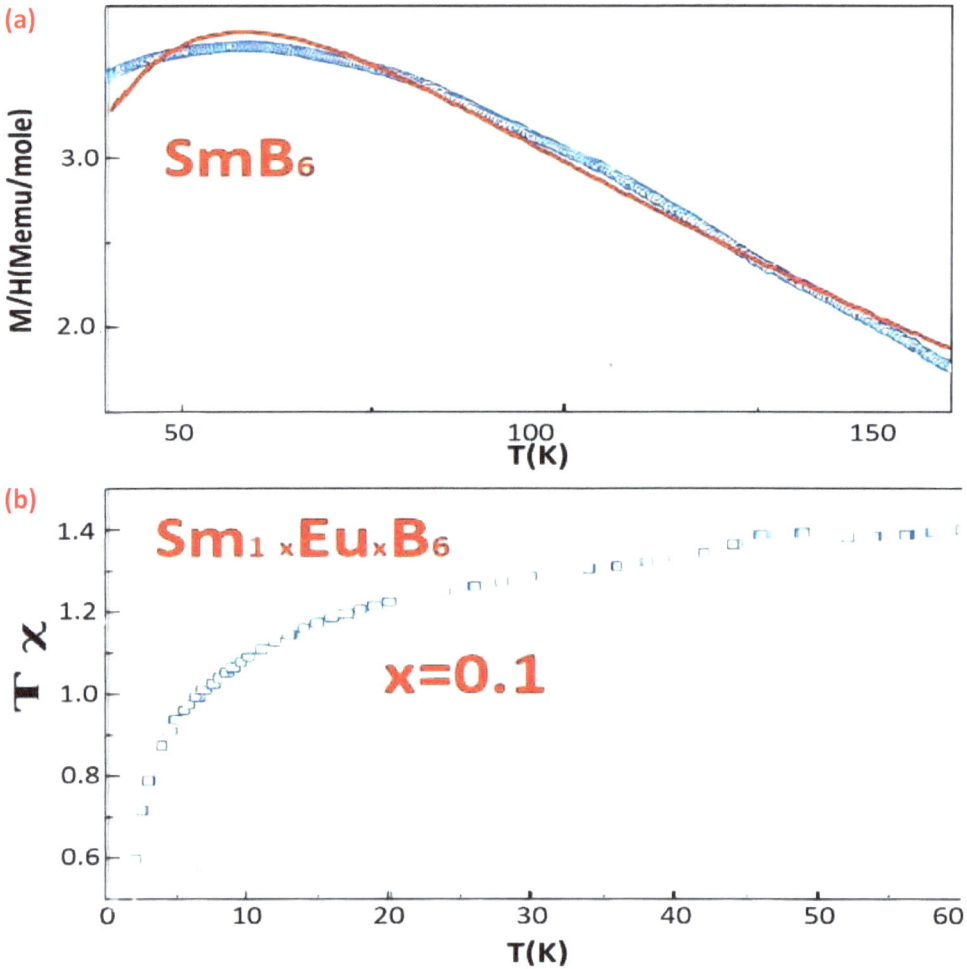

Fig. (5.4). (a) Magnetic sensitivity data between 40 and 150 K for x = 0.0 **(b)** Effective curie constant T χ (T) (x) for x = 0.1 as a function of temperature. Adopted from [23] (Copyright ©2012 American Physical Society).

Otani *et al.* prepared a single large REB_6 crystal (R = La, Ce, Pr, Nd) using a floating zone method [25]. In the study conducted by Li Hong Bao *et al.*, the successful growth of high-quality, large-size triple $La_xCe_{1-x}B_6$ single crystals was done by the floating zone method [26]. The single crystals grown showed excellent field emission and thermionic emission characteristics. The crystal structure, crystal quality, and emission characteristics of $La_xCe_{1-x}B_6$ single crystals were characterized by a 360 Phi scanning single-crystal diffractometer, X-ray Laue diffraction, and thermal electron emission measurements. The field emission single tip of LaB_6 single crystal exhibit an excellent performance than the field emission tip of polycrystalline LaB_6, and its field enhancement factor (β) is 36,231 cm^{-1}. Thus, $La_xCe_{1-x}B_6$ single crystals are used as a hot or cold cathode. Fig. (**5.5a**) shows the quick scan diffraction photo of $La_{0.2}Ce_{0.8}B_6$, which has been magnified at 10 mm/h. However, the results along the directions of hemispherical diffraction (100) show that the crystal produced in two steps consists of twin crystals, as shown in Fig. (**5.5b**). Finally, in Fig. (**5.5c**), the hemispherical diffraction points without the twin structure indicate that a single good-quality crystal of $La_{0.2}Ce_{0.8}B_6$ is obtained.

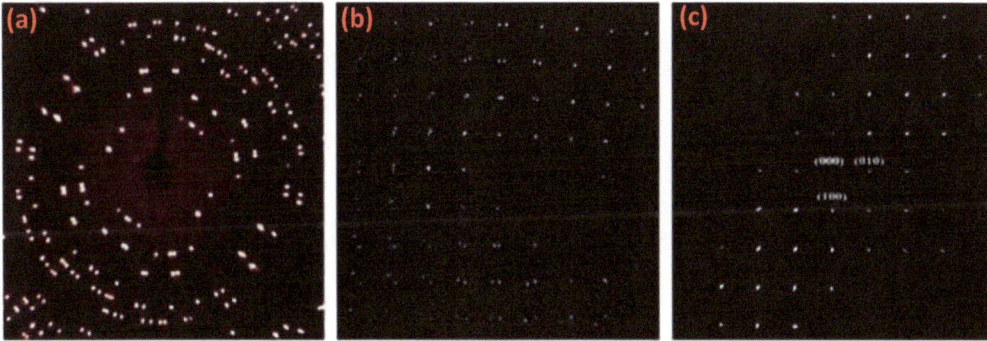

Fig. (5.5). Single crystal diffraction picture of $La_{0.2}Ce_{0.8}B_6$ single crystal, (**a**) for quick scanning, (**b**) and (**c**) for along the [0 0 1] direction grown for second and third time respectively. Adopted from [26] (Copyright © 2012 Elsevier).

LaB_6 single crystal shows superior field emission than multicrystal samples. Therefore, high-quality $La_xCe_{1-x}B_6$ single crystals have excellent thermionic emission or field emission properties. In another study, Rhyee *et al.* found that $Eu_{1-x}La_xB_6$ showed ferromagnetic and metallic properties for $x\leq0,05$ and antiferromagnetic and metallic properties for x = 0,3. Small doping in $Eu_{1-x}La_xB_6$ (x=0,005 and 0,03) exhibits double-pass behavior in transport and magnetic transition maintains a ferromagnetic structure with a constantly saturated moment in magnetization [27]. In another study, Hyun Jung Lee *et al.* measured the thermoelectric power (TEP) of the $Eu_{1-x}La_xB_6$ in the temperature range of $2K\leq T\leq300K$. TEP for EuB_6 (x=0.0 and 0.1), exhibits negative values over the entire temperature range, coherent with the truth that dominant charge carriers on the Fermi surface are electrons. Moreover, for $Eu_{1-x}La_xB_6$ (x=0, and 2), a sign change of the TEP from negative to positive value is shown at about 28 K, which is coherent with the sign reversal of the Hall resistivity of the same alloy [28]. In addition, Nefedova *et al.* measured the magnetic sensitivity, electrical resistance, specific heat, and thermal expansion coefficient of $Sm_{1-x}La_xB_6$ in the temperature range T = 4-300 K. They found that the SmB_6 phonon spectrum of lanthanum doping did not lead to significant changes in anomalies. It has been found that the homogeneous intermediate valent state of the samarium ion is very stable and maintains perfection when the sub-cage Sm is violated [29]. The electronic nature of the REB_6 is responsible for the magnetic sequencing behavior. The metallic REB_6 that exhibits magnetic order is all antiferromagnetic [30]. For example, Glaunsinger *et al.* investigated $Eu_xYb_{1-x}B_6$ exchange interactions at 296 K by the EPR technique. The static magnetic sensitivity studies of the magnetically diluted EuB_6 at the last low temperature have shown that the interaction between europium moments is a continuous ferromagnetic type. Yb is diamagnetic. $Eu_xYb_{1-x}B_6$ for x > 0.2 is powerful paramagnetic at 296 K [31]. In addition, Qidong Li *et al.* successfully synthesized $Pr_xNd_{1-x}B_6$ nanowires at 1030 °C on silicon surfaces by CVD in one step. Qidong Li has shown that semi-aligned nanowires are structurally smooth and well-doped single crystals based on extensive analysis. $Pr_xNd_{1-x}B_6$ nanowires were synthesized in the present study by employing a gas-solid reaction system, and the morphology, structure, and composition of the nanowires were examined. The well-characterized morphology of the $Pr_xNd_{1-x}B_6$ nanowires are quasi-aligned arrays with smooth surfaces and have diameters around 150 nm and lengths extending to more than a few micrometers. The HRTEM results openly show their single-crystalline nature, having terminated facets of indices as (111), (210), and (320) the growth direction of the nanowires have (110) lattice direction. The $Pr_xNd_{1-x}B_6$ nanowires' potential applications contain supplying emitters of electrons for flat panel displays, TEM, SEM, as well as other nanoelectronic building blocks [32]. A different study by Wang *et al.* produced high-quality $La_xCe_yPr_{1-x-y}B_6$ (x = 0.6-0.8, and y = 0.1-0.3) single crystals by the optical floating zone melting method. They studied the crystal structure, mechanical and thermionic emission properties of single crystals growing [33]. The emission current densities increase with the increase in the cathode temperature or applied voltage is shown in Fig. (**5.6**). The maximum emission current density is 90.55 A/cm^2.

Furthermore, in the research conducted by Bao Li-Hong *et al.*, they prepared the $Nd_xGd_{1-x}B_6$ alloy using the mixed powder of GdH_2, NdH_2 and B using the reactive spark plasma sintering method. They examined the effects of Nd doping on the crystal structure, grain orientation, thermionic emissions, and magnetic properties of the hexaboride. All samples are sintered by the SPS method, and the samples exhibit high density (>95%) and high Vickers hardness values (2319 kg /mm^2) [34].

In the REB_6 family, LaB_6 and PrB_6 have low working functions (LaB_6 ~ 2,6 eV [35] and PrB_6 ~ 3,1 eV) [36], electron field emission (FE) materials used in electron microscopy and electron emitter production due to low volatility and high brightness. In a different study, Wei Khan *et al.* manufactured Lantan-praseodymium hexaboride ($La_xPr_{1-x}B_6$) nanoparticles using a simple flux-controlled method using lanthanum (La) powders, praseodymium (Pr) powders, and bor chloride (BCl_3) gas as starting materials at 1050 °C. Scanning electron microscopic (SEM) found that the nanoparticles had a length of 2-10 μm and a diameter of 50 to 300 nm at the roots and 10-80 nm at the tips [37].

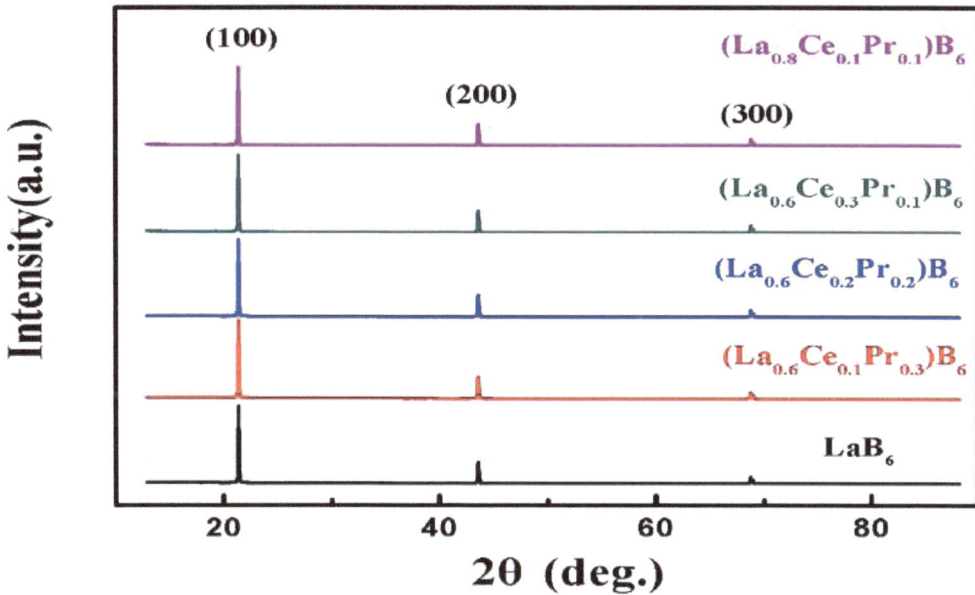

Fig. (5.6). Thermionic emission current densities of $La_{0.6}Ce_{0.1}Pr_{0.3}B_6$ single crystals. Adopted from [33] (Copyright © 2018 Elsevier B.V.).

CONCLUSION

Alloying REB_6 with other metals improves the properties of REB_6. For example, the alloying reduces the total kinetic energy of the electrons of REB_6 and provides the absorption valleys altering to longer wavelength direction and the red-shift in the absorption valley [3]. The alloying of REB_6 can improve DOS and the position of the Fermi energy level. Moreover, the work function of the alloyed REB_6 can be noticeably decreased, meaning better emission performance than REB_6. Furthermore, alloying of REB_6 leads to a reduction of the plasmon energy and plasma frequency of REB_6, leading to better optical performance.

REFERENCES

[1]   Callister, W.D., Jr; Rethwisch, D.G. *Callister's materials science and engineering*; John Wiley & Sons, **2020**.

[2]   Dossett, J.L.; Boyer, H.E. *Practical heat treating*; Asm International, **2006**.
      [http://dx.doi.org/10.31399/asm.tb.pht2.9781627082624]

[3]   Qi, X.; Bao, L.; Chao, L.; Tegus, O. Experimental and theoretical investigation on the tunable optical property of nanocrystalline Ca-doped CeB_6; **2018**.
      [http://dx.doi.org/10.1016/j.physb.2017.12.003]

[4]   Liu, H.; Zhang, X.; Xiao, Y.; Zhang, J. The electronic structures and work functions of (100) surface of typical binary and doped REB_6 single crystals. *Appl. Surf. Sci.,* **2018**, *434*, 613-619.
        [http://dx.doi.org/10.1016/j.apsusc.2017.10.233]

[5]   Chao, L.; Bao, L.; Wei, W.; Tegus, O. Optical properties of Yb-doped LaB_6 from first-principles calculation. *Mod. Phys. Lett. B,* **2016**, *30*(7), 1650091.
        [http://dx.doi.org/10.1142/S0217984916500913]

[6]   Huang, B.; Duan, Y.H.; Sun, Y.; Peng, M.J.; Chen, S. Electronic structures, mechanical and thermodynamic properties of cubic alkaline-earth hexaborides from first principles calculations. *J. Alloys Compd.,* **2015**, *635*, 213-224.
        [http://dx.doi.org/10.1016/j.jallcom.2015.02.128]

[7]   Otani, S.; Honma, S.; Ishizawa, Y. Preparation of LaB_6 single crystals by the floating zone method. *J. Alloys Compd.,* **1993**, *193*(1-2), 286-288.
        [http://dx.doi.org/10.1016/0925-8388(93)90373-U]

[8]   Kuznetsov, G.I.; Sokolovsky, E.A. Dependence of effective work function for LaB_6 on surface conditions. *Phys. Scr.,* **1997**, *T71*(T71), 130-133.
        [http://dx.doi.org/10.1088/0031-8949/1997/T71/024]

[9]   Schelm, S.; Smith, G.B. Dilute LaB_6 nanoparticles in polymer as optimized clear solar control glazing. *Appl. Phys. Lett.,* **2003**, *82*(24), 4346-4348.
        [http://dx.doi.org/10.1063/1.1584092]

[10]   Takeda, H.; Kuno, H.; Adachi, K. Solar control dispersions and coatings with rare-earth hexaboride nanoparticles. *J. Am. Ceram. Soc.,* **2008**, *91*(9), 2897-2902.
        [http://dx.doi.org/10.1111/j.1551-2916.2008.02512.x]

[11]   Bao, L.; Qi, X.; Tana, T.; Chao, L.; Tegus, O. Effects of induced optical tunable and ferromagnetic behaviors of Ba doped nanocrystalline LaB_6. *Phys. Chem. Chem. Phys.,* **2016**, *18*(28), 19165-19172.
        [http://dx.doi.org/10.1039/C6CP03022J] [PMID: 27362626]

[12]   Chao, L-M.; Bao, L-H.; O, T. Optical response of CeB_6 nanoparticles with different sizes and shapes from discrete-dipole approximation. *Chin. Phys. Lett.,* **2015**, *32*(4), 043301.
        [http://dx.doi.org/10.1088/0256-307X/32/4/043301]

[13]   Qi, X.; Bao, L.; Chao, L.; Tegus, O. Experimental and theoretical investigation on tunable optical property of nanocrystalline Ca-doped CeB_6. *Physica B,* **2018**, *530*, 312-316.
        [http://dx.doi.org/10.1016/j.physb.2017.12.003]

[14]   Uimin, G. Quadrupolar and magnetic ordering in CeB_6. *Phys. Rev. B Condens. Matter,* **1997**, *55*(13), 8267-8279.
        [http://dx.doi.org/10.1103/PhysRevB.55.8267]

[15]   Fisk, Z.; Johnston, D.C.; Cornut, B.; von Molnar, S.; Oseroff, S.; Calvo, R. Magnetic, transport, and thermal properties of ferromagnetic EuB_6. *J. Appl. Phys.,* **1979**, *50*(B3), 1911-1913.
        [http://dx.doi.org/10.1063/1.327162]

[16]   Henggeler, W.; Ott, H.R.; Young, D.P.; Fisk, Z. Magnetic ordering in EuB_6, investigated by neutron diffraction. *Solid State Commun.,* **1998**, *108*(12), 929-932.
        [http://dx.doi.org/10.1016/S0038-1098(98)00470-0]

[17]   Rhyee, J.S.; Oh, B.H.; Cho, B.K.; Kim, H.C.; Jung, M.H. Magnetic properties in Ca-doped Eu hexaborides. *Phys. Rev. B Condens. Matter,* **2003**, *67*(21), 212407.
        [http://dx.doi.org/10.1103/PhysRevB.67.212407]

[18]   Batkova, M.; Batko, I.; Bauer, E.; Khan, R.T.; Filipov, V.B.; Konovalova, E.S. Effect of pressure on the electric transport properties of carbon-doped. *Solid State Commun.,* **2010**, *150*(13-14), 652-654.
        [http://dx.doi.org/10.1016/j.ssc.2009.12.025]

[19]   Dzero, M.; Sun, K.; Galitski, V.; Coleman, P. Topological Kondo insulators. *Phys. Rev. Lett.,* **2010**,

104(10), 106408.
[http://dx.doi.org/10.1103/PhysRevLett.104.106408] [PMID: 20366446]

[20]   Ivashchenko, V.I.; Turchi, P.E.A.; Shevchenko, V.I.; Medukh, N.R.; Leszczynski, J.; Gorb, L. Electronic, thermodynamics and mechanical properties of LaB_6 from first-principles. *Physica B,* **2018,** *531,* 216-222.
[http://dx.doi.org/10.1016/j.physb.2017.12.044]

[21]   Mattox, T.; Urban, J. Tuning the Surface Plasmon Resonance of Lanthanum Hexaboride to Absorb Solar Heat: A Review. *Materials (Basel),* **2018,** *11*(12), 2473.
[http://dx.doi.org/10.3390/ma11122473] [PMID: 30563148]

[22]   Bao, L.; Ning, J.; Liu, Z. Mechanism for Transmittance Light Tunable Property of Nanocrystalline Eu-doped SmB_6: Experimental and First-Principles Study. *J. Rare-Earths,* **2020,** *39*(9), 1100-1107.

[23]   Yeo, S.; Song, K.; Hur, N.; Fisk, Z.; Schlottmann, P. Effects of Eu doping on SmB_6 single crystals. *Phys. Rev. B Condens. Matter Mater. Phys.,* **2012,** *85*(11), 115125.
[http://dx.doi.org/10.1103/PhysRevB.85.115125]

[24]   Akintola, K.; Pal, A.; Potma, M.; Saha, S.R.; Wang, X.F.; Paglione, J.; Sonier, J.E. Quantum spin fluctuations in the bulk insulating state of pure and Fe-doped SmB_6. *Phys. Rev. B,* **2017,** *95*(24), 245107.
[http://dx.doi.org/10.1103/PhysRevB.95.245107]

[25]   Otani, S.; Nakagawa, H.; Nishi, Y.; Kieda, N. Floating zone growth and high temperature hardness of rare-earth hexaboride crystals: LaB_6, CeB_6, PrB_6, NdB_6, and SmB_6. *J. Solid State Chem.,* **2000,** *154*(1), 238-241.
[http://dx.doi.org/10.1006/jssc.2000.8842]

[26]   Bao, L.H.; Tegus, O.; Zhang, J.X.; Zhang, X.; Huang, Y.K. Large emission current density of $La_xCe_{1-x}B_6$ high quality single crystals grown by floating zone technique. *J. Alloys Compd.,* **2013,** *558,* 39-43.
[http://dx.doi.org/10.1016/j.jallcom.2012.11.098]

[27]   Rhyee, J.S.; Kim, C.A.; Cho, B.K.; Ri, H.C. Electron doping dependence of ferromagnetism in $Eu_{1-x}La_xB_6$. *Phys. Rev. B Condens. Matter,* **2002,** *65*(20), 205112.
[http://dx.doi.org/10.1103/PhysRevB.65.205112]

[28]   Lee, H.J.; Kim, M-H.; Park, S.H.; Kim, H.C.; Kim, J.Y.; Cho, B.K. Thermoelectric power study of $Eu_{1-x}La_xB_6$ (x= 0.0, 0.1, 0.2). *Physica B,* **2006,** *378,* 626-627.
[http://dx.doi.org/10.1016/j.physb.2006.01.180]

[29]   Nefedova, E.V.; Alekseev, P.A.; Klement'ev, E.S.; Lazukov, V.N.; Sadikov, I.P.; Khlopkin, M.N.; Tsetlin, M.B.; Konovalova, E.S.; Paderno, Y.B. Imperfection of the Sm sublattice and valence instability in compounds based on SmB_6. *J. Exp. Theor. Phys.,* **1999,** *88*(3), 565-573.
[http://dx.doi.org/10.1134/1.558830]

[30]   Matthias, B. T.; Geballe, T. H.; Andres, K.; Corenzwit, E.; Hull, G. W.; Maita, J. P. Superconductivity and antiferromagnetism in boron-rich lattices. *Science (80-.).,* **1968,** *159*(3814), 530.
[http://dx.doi.org/10.1126/science.159.3814.530]

[31]   Glaunsinger, W.S. EPR study of exchange interactions in $Eu_xYb_{1-x}B_6$. *J. Magn. Reson.,* **1975,** *18*(2), 265-275.

[32]   Li, Q.; Zhao, Y.; Fan, Q.; Han, W. A one-step chemical vapor deposition approach for solid solution $Pr_x Nd_{1-x}B_6$ nanowire growth. *RSC Advances,* **2015,** *5*(117), 96412-96415.
[http://dx.doi.org/10.1039/C5RA19820H]

[33]   Wang, Y.; Zhang, J.; Yang, X.; Zhu, Z.; Zhao, J.; Xu, B.; Li, Z. High-quality $La_xCe_yPr_{1-x-y}B_6$ single crystal with excellent thermionic emission properties grown by optical floating zone melting method. *J. Alloys Compd.,* **2018,** *769,* 706-712.
[http://dx.doi.org/10.1016/j.jallcom.2018.08.046]

[34]   Bao, L.-H.; Zhang, J-X.; Zhou, S-L.; Tegus, Synthesis, thermionic emission and magnetic properties of
       (Nd$_x$Gd$_{1-x}$)B$_6$. *Chin. Phys. B,* **2011**, *20*(5), 058101.
       [http://dx.doi.org/10.1088/1674-1056/20/5/058101]

[35]   Yutani, A.; Kobayashi, A.; Kinbara, A. Work functions of thin LaB$_6$ films. *Appl. Surf. Sci.,* **1993**, *70-
       71*, 737-741.
       [http://dx.doi.org/10.1016/0169-4332(93)90612-F]

[36]   Decker, R.W.; Stebbins, D.W. Photoelectric work functions of the borides of lanthanum,
       praseodymium, and neodymium. *J. Appl. Phys.,* **1955**, *26*(8), 1004-1006.
       [http://dx.doi.org/10.1063/1.1722123]

[37]   Han, W.; Zhang, H.; Chen, J.; Zhao, Y.; Fan, Q.; Li, Q.; Liu, X.; Lin, X. Single-crystalline La Pr$_1$−B$_6$
       nanoawls: Synthesis, characterization and growth mechanism. *Ceram. Int.,* **2016**, *42*(5), 6236-6243.
       [http://dx.doi.org/10.1016/j.ceramint.2016.01.006]

CHAPTER 6

The Rare-Earth Hexaboride Based Composites

Abstract: Rare-Earth metal hexaborides (REB_6) can be composited with some kind of ceramics, such as SiC, MgO, Carbon Nanotube, and Alumina. These types of composites can show excellent mechanical, optical, and thermionic properties. For example, SiC ceramics have high condensation behavior, high corrosion resistance, high thermal shock resistance, and high hardness properties; MgO ceramics have high fire resistance, high thermal conductivity, and low electrical conductivity properties; Carbon nanotubes have high optical and mechanical properties and Al_2O_3 ceramics have high abrasion and corrosion resistance and low density. The sizes of these materials are also significant as nano, and micro-sized ceramic materials have different properties when forming a composite with REB_6 or any materials.

Keywords: Mechanical properties, High friction resistance, High thermal shock resistance, Microstructure.

6.1. INTRODUCTION

Due to their high melting points, excellent strength, and creep resistance, REB_6-based composites can be used as a good structural material at high temperatures (>1200°C) [1, 2]. REB_6 structures are composited with some kind of ceramics, such as $X^{IV}B_2$ (X^{IV}-Ti, Zr, Rf, and Hf), SiC, MgO, Carbon Nanotube, Alumina, *etc.* They indicate important properties, such as strong hardness and friction resistance at high temperatures, making them attractive as structural materials. For example, $X^{IV}B_2$ (X^{IV} - Ti, Zr, Rf, and Hf) ceramics have high hardness, bending strength, and stress hardening at high temperatures; SiC ceramics have high condensation behavior, high corrosion resistance, high thermal shock resistance, high hardness properties; MgO ceramics have high fire resistance, high thermal conductivity, and low electrical conductivity properties; CNT materials have high optical and mechanical properties and Al_2O_3 ceramics have high abrasion and corrosion resistance and low density. The sizes of these materials are also important since nano and micro-sized ceramic materials show different properties when forming a composite with REB_6 or any materials.

Mikail Aslan and Cengiz Bozada

There are many available methods for the production of composites consisting of REB_6 (see Chapter 4). For example, LaB_6-ZrB_2 composites are produced by vacuum hot press sintering technique and prepared by floating zone method based on melting of the crucible free zone of powders. LaB_6-TiB_2 composites are produced by the floating zone method. The method is based on crucible-free zone melting, and the products are obtained by dissolving them in argon flow in the electric arc. CeB_6-$TiSi_2$ composites are produced by sintering with reactive spark plasma (SPS). LaB_6-MgO composites are produced by the screen printing technique and magnetron spraying method. LaB_6-CNT and CeB_6-CNT composites are synthesized by physical (PVD) and chemical vapor deposition (CVD). LaB_6-Al_2O_3 composites are produced with high-energy ball grinding, annealing, and leaching processes. After synthesizing these composites, hardness, surface morphology, optical, discharge properties, fracture strength, condensation behavior, microstructure, and mechanical properties of REB_6 composites can be analyzed by TEM, XRD, SEM, Raman, and XPS.

In this part, we have focused on phase stability, microstructure, mechanical performance, and absorption properties of composites consisting of nano and micro ceramic materials with REB_6.

6.2. REB_6-$XIVB_2$ COMPOSITES

REB_6-$X^{IV}B_2$ (X^{IV}=Ti, Zr, Hf, or V) composites have low work functions, high thermionic current and electron emission density, mechanical properties, and enhanced thermal-shock resistance [3]. Due to these properties, REB_6-$X^{IV}B_2$ composites are one of the best candidates for electrode application in space propulsion [4]. $X^{IV}B_2$ has a higher melting point and Young's modulus than REB_6 single crystal [5]. REB_6-$X^{IV}B_2$ has a large improvement in current density compared to single-crystal REB_6. Furthermore, REB_6-$X^{IV}B_2$ composites exhibit improved thermal cycling reliability as compared with REB_6 [6].

Due to its high melting point, REB_6-$X^{IV}B_2$ can also be used as a good structural material at high temperatures. Existing oxide-oriented solidified composites have attracted much attention in this field because they exhibit excellent strength and frictional resistance at high temperatures (> 1200°C), making them attractive as structural materials [7, 8]. However, in this class of materials, the fracture strength is low since the interfaces between the two phases typically adopt low-energy bonding orientation relationships during the directional solidification process, which supports strong bonding and prevents the sticking of the interfaces [9]. Directly crystallized composites of REB_6-$X^{IV}B_2$ (X^{IV}-Ti, Zr, Rf, and Hf) are the most investigated ceramic materials [10 - 12].

6.2.1. LaB$_6$-ZrB$_2$ Composites

Zirconium diborides (ZrB$_2$) are grey, have a very high melting temperature (3245°C), and their crystal structures are hexagonal. ZrB$_2$, carbon, zirconium, and boron oxide can be produced by reacting in the electric arc furnace. Furthermore, ZrB$_2$ has oxidation resistance, high hardness, and thermal shock resistance. ZrB$_2$ ceramics are used as molten metal containers, diffusion barriers in semiconductors, and ignite absorbers in nuclear reactor cores [13].

Lanthanum hexaborides(LaB$_6$) have low work functions. LaB$_6$-ZrB$_2$ composites have a current density of 4-5 times higher than that of pure LaB$_6$. The resistance of LaB$_6$-ZrB$_2$ composites to thermal shock and poisoning is higher than that of pure LaB$_6$ [14]. Paderno *et al.* reported that the boron-boron (B-B) distance in LaB$_6$ can be modified by adding ZrB$_2$. They suggested that if the B-B distance in the hexaboride was close to the distance in the diboride, the semi-compliant interface could be formed between the LaB$_6$-ZrB$_2$ composite [11]. In a different study, Chen *et al.* prepared LaB$_6$-ZrB$_2$ composites from LaB$_6$ and ZrB$_2$ powders, then pressed and melted them with an electric arc to form rod samples. They performed directional solidification in a vacuum heated at 2900 ° C with the electron beam and floating zone melting furnace [12]. In a separate study, Wang *et al.* characterized the microstructures of LaB$_6$-ZrB$_2$ composites with transmission electron microscopes (TEM, 100CXII or HRTEM, TECNAI F-30, Philips, Holland) equipped with energy-separated X-ray spectrometry (EDS). LaB$_6$-ZrB$_2$ composites were able to magnify along one direction to show well-oriented ZrB$_2$ fiber and well-dispersed LaB$_6$ matrix after directional solidification [15]. In a different study, Min *et al.* produced polycrystalline LaB$_6$-ZrB$_2$ composites with different ZrB$_2$ content by vacuum hot press sintering technique 330.0MPa and 3.70 MPa·m$^{1/2}$ according to the results obtained, the hardness and flexural strength of LaB$_6$-ZrB$_2$ polycrystalline increased with increasing ZrB$_2$ content; however, fracture toughness first reaches a peak corresponding to a content of ZrB$_2$ of 21% by weight. In the microstructure observation, a concentration was detected due to the addition of ZrB$_2$. Fig. (**6.1**) shows the detailed morphology of transgranular fracture in the LaB$_6$ matrix and intergranular in the ZrB$_2$ field [16]. In a similar study, Bogomol *et al.* prepared a directed LaB$_6$-ZrB$_2$ by floating zone method based on melting of the crucible free zone of compressed powders. ZrB$_2$ and LaB$_6$ powders were used as the first materials. The flexural strength of the composite was evaluated in the temperature range of 25-1600 ˚C and reached 950 MPa at 1600 ˚C. The fracture strength, SEM, and TEM hardening mechanisms were investigated under different conditions. They predicted that the strength of LaB$_6$-ZrB$_2$ at 25-1200 ˚C was mainly associated with crack deflection, bridge hardening mechanisms, increased plasticity of the ZrB$_2$

phase, and increased plasticity of matrices and fibers at 1200-1600 ˚C. The dislocated structure of the fibers was analyzed, revealing the formation of stress during high-temperature deformation in the single crystal ZrB_2. SEM-EDS analysis of a polished surface of the directionally solidified LaB_6-ZrB_2 composite revealed that it consisted of a LaB_6 matrix reinforced with ZrB_2 fiber (Fig. **6.1a**). A representative EDS spectrum (Fig. **6.1b**) shows B peaks at high energies and a strong signal from Zr and La [1].

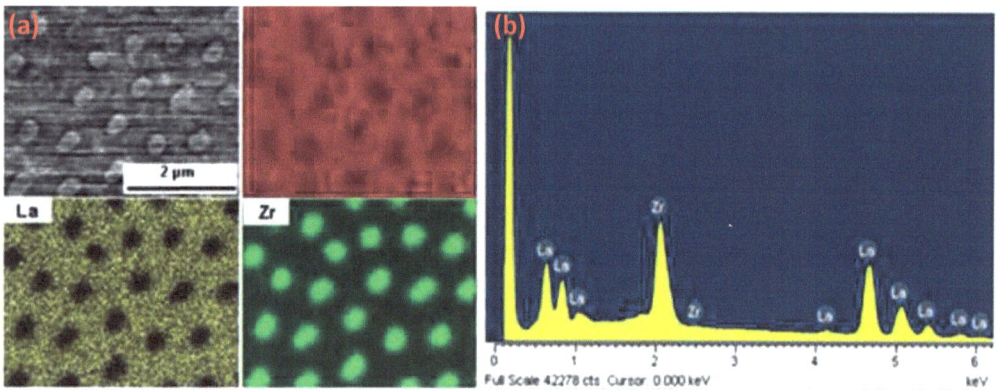

Fig. (6.1). (a) Mapping of the directed LaB_6-ZrB_2 composite and **(b)** EDS spectra. Adopted from [1] (Copyright © 2011 Elsevier BV All rights reserved).

Gao *et al.* studied the oxidation behavior of LaB_6-ZrB_2 composites in the temperature range of 600-1300 ° C. LaB_6-ZrB_2 composites were sintered by hot pressing. They found that the weight of the material increased as the oxidation temperature and the holding time increased [17]. SEM of the surfaces of oxidized LaB_6-ZrB_2 at dissimilar oxidation conditions is shown in Fig. (**6.2**). The roughness of some parts of the surface when the sample is exposed to air at 1000 ° C for 6 hours. The oxidation resistance of LaB_6-ZrB_2 composites is higher than that of monolithic LaB_6.

6.2.2. LaB_6-TiB_2 Composites

TiB_2 has a high melting temperature, high abrasion resistance, high hardness, high strength, and high resistance to chemicals. It is resistant to sintering and is usually densified by isostatic pressing or hot press. High purity can be achieved by non-pressurized sintering of TiB_2, but auxiliaries such as carbon, chromium, and iron are needed in liquid form [18]. The thermal expansion coefficient of LaB_6 is smaller than TiB_2 at all temperatures [19].

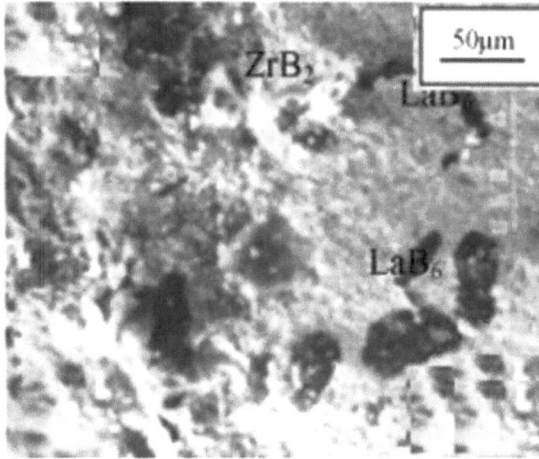

Fig. (6.2). Microstructures of LaB_6-ZrB_2 eutectic composite surfaces oxidized at 1000 ° C. Adopted from [17] (Copyright © 2004 Techna S.r.l. Published by Elsevier Ltd.).

The LaB_6-TiB_2 composite exhibits very high bending strength. Bogomol *et al.* prepared a guided solidified LaB_6-TiB_2 composite with a floating zone method based on the melting of the pot-free zone of compressed dust. When they examined its structure as the first material, they revealed monocrystalline TiB_2 stress hardening during high-temperature deformation [20]. In Fig. (**6.3a**), the stretched fibers are divided into regions with different deflection lines. Thus, the displacement of the monocrystalline TiB_2 was observed during high-temperature deformation. In region I, near the matrix phase, displacement lines are directed towards the fiber axis corresponding to the shear plane of the crystalline structure (0001) of the hexagonal $TiB_2(P_6)$. II. the deformation rate in the region is higher, and the stress is less homogeneous. III. The displacement density near the fiber fracture in the region increases to 10^9 cm$^{-2,}$ and dispersed cellular infrastructure is observed. Fig. (**6.3b**) shows stress-free TiB_2 fibers in the directly reinforced LaB_6-TiB_2 composite. TiB_2 and LaB_6 powders were used. The bending strength of the molten LaB_6-TiB_2 composite was evaluated in the temperature range of 1000-1600 °C and reached 470 MPa at 1400 °C. They estimated that the bending strength of the directionally reinforced LaB_6-TiB_2 composite at high temperatures depended mainly on the plasticity of the TiB_2 fibers and the LaB_6 matrix.

In another study, Soloviova *et al.* obtained polycrystalline (PC) and LaB_6-TiB_2, PC, and (111) powders used in single-crystalline (MC) seeds based on the floating zone melting method. This study aimed to determine the thermal annealing, residual stresses, and mechanical properties of polycrystalline LaB_6-TiB_2 ceramic composites at a temperature of 2370/2408 °C [21]. In a different study, Nesmelov *et al.* investigated the microstructure, hardness, and fracture strength of the electric arc-melted LaB_6-TiB_2 composite. In contrast to the floating zone method, they found that arc melting did not allow for a unidirectional solidifying alloy on a macro scale. The structure of arc-melted alloys can be formed by the different orientations of the blocks (according to different temperatures during crystallization). They obtained the LaB_6-TiB_2 alloy, which solidifies in a direction by melting it in the argon flow in the electric arc. The structure of the composites is made up of TiB_2 fibers (rods), which are directionally crystallized in the LaB_6 matrix [2].

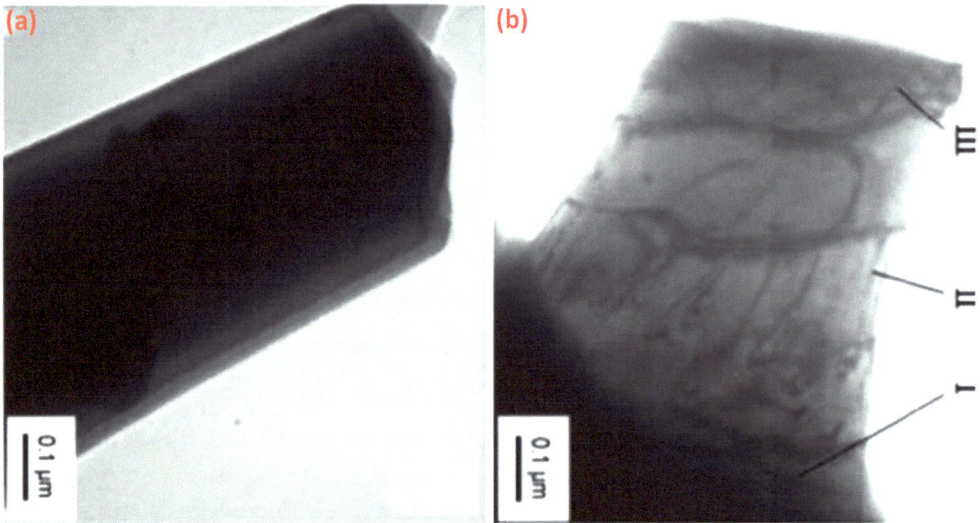

Fig. (6.3). TEM images of **(a)** unstretched TiB_2 fibers at 1400 °C and **(b)** directly reinforced LaB_6-TiB_2 composite of stretched TiB_2 fibers. Adopted from [20] sevier BV All rights reserved).

6.2.3. CeB_6-$TiSi_2$ Composites

Titanium disilicide ($TiSi_2$) is widely used in the semiconductor industry due to its low resistance, high-temperature stability, and chemical compatibility [22]. The melting point, hardness, elastic modulus, density, and electrical resistivity of $TiSi_2$ are lower than CeB_6. Sonber *et al.* synthesized CeB_6 powder by boron carbide reduction of CeO_2 in a vacuum at 1600 ° C. They used $TiSi_2$ as a sintering additive due to its improved density and lower melting point during the sintering of CeB_6. This study includes the results of consolidation with SPS and the investigation of mechanical properties [23].

6.2.4. GdB_6-TiB_2 Composites

Titanium diboride (TiB_2) is an extremely significant advanced ceramic material due to its excellent features, and good thermal, including outstanding hardness, extreme wear resistance, high melting point, electrical conductivity, and chemical inertness [24]. GdB_6 has attracted more effort and interest due to a lot of physical properties of importance for science and technology. For example, due to its good work function, magnetic transport properties, field emission properties, and electrical and optical properties, GdB_6 is an excellent infrared absorption/reflective material that can be used as a solar radiation protective material for visible light permeability high windows [25]. The melting point of TiB_2 is bigger than GdB_6. GdB_6-TiB_2 composites don't mix up to a temperature of 2670. Nikolaeva *et al.* synthesized GdB_6-TiB_2 composites and examined the wide concentration and temperature ranges of interaction in the GdB_6-TiB_2 system. The powders of produced GdB_6-TiB_2 were used by the Donetsk Plant of Chemical Reagents. X-ray phase analysis of GdB_6 powder annealed at 17700K in a vacuum for 4 h showed no additions of gadolinium tetraboride [26].

6.2.5. GdB_6-HfB_2 Composites

Hafnium diboride (HfB_2) belongs to ultra-high-temperature ceramic glass, a kind of ceramic consisting of hafnium and boron. The melting temperature is about 3250 °C. It is an unusual ceramic with relatively high thermal and electrical conductivity, with features shared with zirconium diboride and isostructural titanium diboride. It is a grey, metallic-looking material.

HfB_2 has a hexagonal crystal structure, a density of 10.5 grams per cubic centimeter, and a molar mass of 200.11 grams per mole [27]. The work on the GdB_6-HfB_2 composite is important. Especially thermal and measurement of microhardness. Nikolaeva *et al.* produced GdB_6-HfB_2 composites. They investigated GdB_6-HfB_2 composites by X-ray phase analyses and thermal measurement of microhardness. The pseudobinary vertical section of the GdB_6-HfB_2 system is shown in Fig. **(6.4)**. In identifying the pseudobinary system (a vertical cut from the Gd-Hf-B ternary system, where GdB_6 melts inharmoniously), 26 experimental specimens of 13 compositions in 5 or 10 wt% increments across the binary were used. The specimens were sintered in an electric furnace at 1797 °C for 1 h under high-purity argon [28].

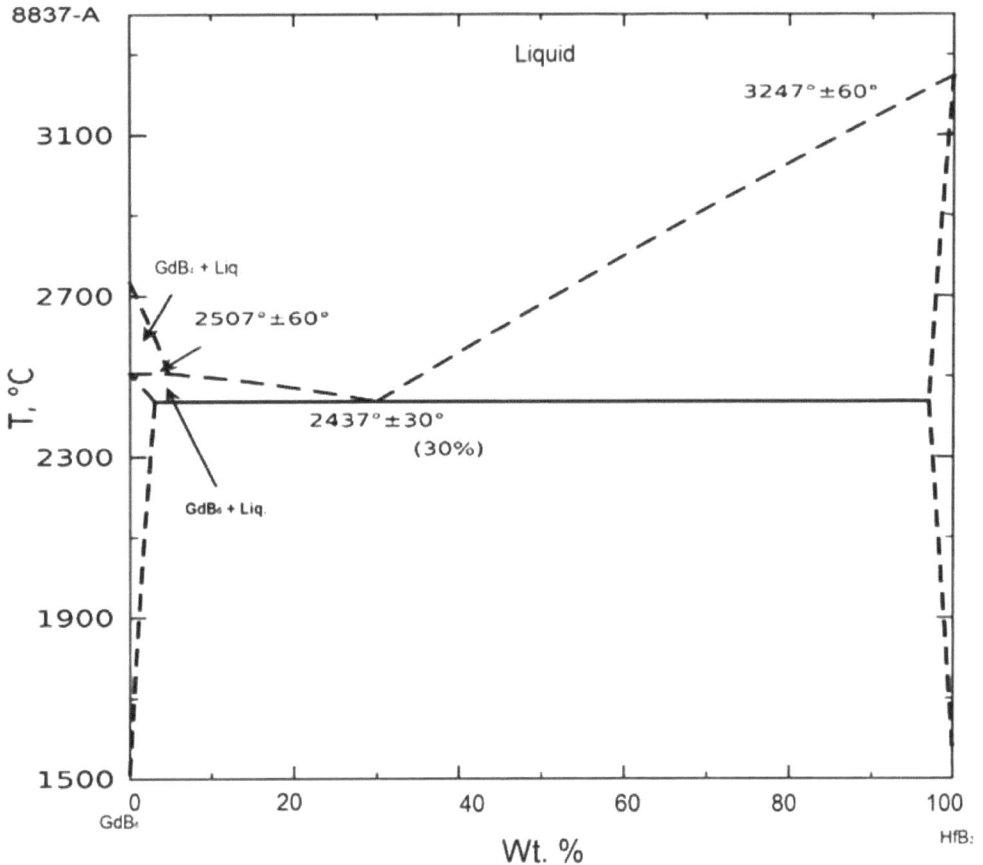

Fig. (6.4). The pseudobinary vertical section of the GdB_6-HfB_2 system [28] (Copyright © 2020 International Atomic Energy Agency (IAEA)).

6.3. LAB$_6$-SIC COMPOSITES

SiC ceramics have high young modulus (441- 475 GPa), high melting temperature (2545 °C), high hardness (24.5-28.2 GPa), high oxidation, low theoretical density (3.21 g / cm3) features such as corrosion resistance, high thermal shock resistance, low thermal expansion coefficient, low thermal conductivity, allow them to find many applications in materials such as aerospace industry, polishing processes and abrasive tool materials, advanced technology ceramics and armor materials [29].

LaB$_6$ -SiC composite is attractive because it combines the excellent mechanical performance of SiC with the high electron emission efficiency of LaB$_6$ [30, 31]. On the other hand, due to the high melting temperature, strong covalent bond, and low self-diffusion coefficients of SiC-LaB$_6$ composites, it is difficult to prepare a high-density compound using conventional sintering without any additives [32, 33]. Yang *et al.* prepared a SiC-LaB$_6$ composite with the SPS technique and systematically examined the condensation behavior, microstructure properties, and mechanical properties. The SPS technique prepared a SiC-LaB$_6$ composite at a pressure of 40 MPa for 300 seconds at 1600-1880 °C and investigated the condensation behavior, microstructure properties, and mechanical properties. Sintered SiC-LaB$_6$ at 1840 °C found a partial density of 98.6% max. The maximum Vickers hardness and fracture toughness obtained for the sintered SiC-LaB$_6$ composite at 1840 ° C were 26.5 GPa and 7.15 MPa m$^{1/2}$, respectively. They concluded that crack collapse and crack branching play an important role in the hardening of the SiC-LaB$_6$ composite. Fig. **(6.5)** shows the XRD analysis of the mixing powders after milling and the sintered SiC-LaB$_6$ composite at 1600-1880 ° C for 3h at 40 MPa pressure. No shifts were detected at SiC or LaB$_6$ peaks, indicating that no chemical reaction occurred between these two components at 1600-1880 °C [34].

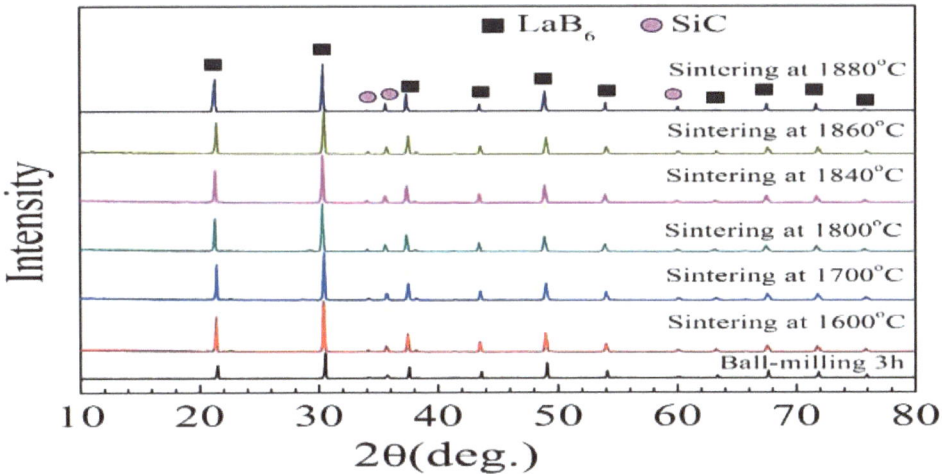

Fig. (6.5). XRD patterns of sintered SiC-LaB$_6$ composite at 1600-1880 ° C and powder mixtures after ball grinding 3 hours. Adopted from [34] (Copyright © 2017 Elsevier BV All rights reserved).

6.4. LAB$_6$-MGO COMPOSITES

Magnesium oxide (MgO) is found in crystal and white colors [35]. The important feature of MgO is its high thermal conductivity, low electrical conductivity, and fire resistance. MgO material is used as a protective layer due to its high durability and good protection against ion bombardment. In the alternating current plasma display panel (AC PDP),

LaB$_6$-MgO composites are used [36]. LaB$_6$ thin films have much lower firing voltages and shorter discharge delay times than pure MgO thin films. Also, LaB$_6$-MgO ensures high enough transmittance. Deng *et al.* prepared LaB$_6$ nanofilm composites on the MgO layer using the screen printing technique to create a new AC PDP protective layer. Compared to pure MgO film, the test panel with a printed LaB$_6$/MgO protective layer showed better discharge performance than pure MgO nanocomposite film [37]. In another study carried out by Deng *et al.* The magnetron sputtering method was used, which is compatible with the manufacturing processes of AC PDP to produce LaB$_6$ doped MgO protective layers, and examined their discharge performance in AC PDP test panels. They studied surface morphology, optical properties, and discharge characteristics with varying concentrations of LaB$_6$ doping. This nanocomposite film is effective in reducing the ignition voltage of AC PDP. In particular, when the doping concentration of LaB$_6$ is 3%, the ignition voltage is reduced by 17% below 10% Xe-Ne of 400 Torr compared to the conventional MgO layer. Fig. **(6.6)** shows the transmission spectrum of LaB$_6$ doped MgO / MgO double protective layer nanocomposite films using the conventional MgO protective layer as a reference. It has been clearly shown that the permeability of the new protective layer is gradually reduced with the increase in LaB$_6$ doping concentration. When the mass ratio increases from 1% to 10%, transmission is reduced by 92-95% and 76-82%. There are two possible causes of high transmission reduction in the LaB$_6$ mass ratio: LaB$_6$ is opacity and surface morphology [38].

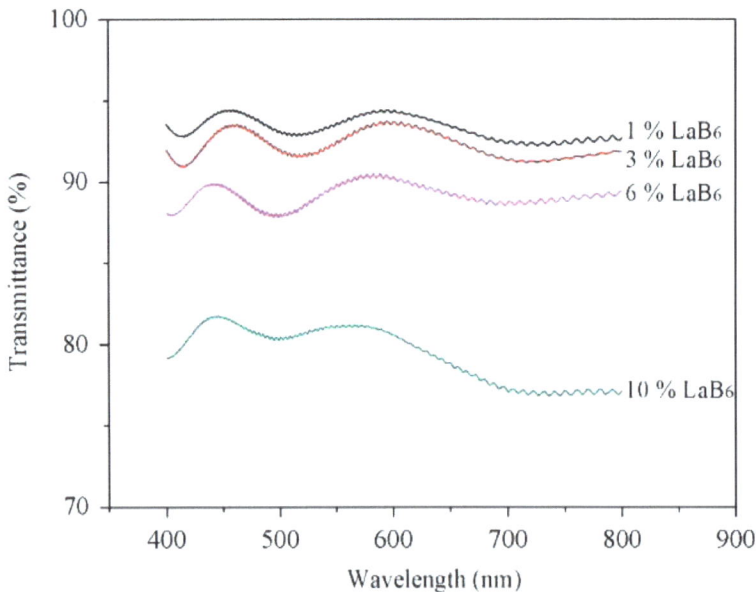

Fig. (6.6). Permeability of MgO/ MgO protective layers with LaB$_6$. Adopted from [38] (Copyright © 2014 Elsevier BV All rights reserved).

6.5. LAB$_6$-SIO$_2$ COMPOSITES

Silicon dioxide or silica (SiO_2) is a compound of Silicon and Oxygen, commonly known as silica, and the elements are linked by a covalent bond. It is one of the components of the sand and can be found naturally in Quartz. Silica is one of the most complicated and plentiful families of materials, found as a compound of several minerals and a synthetic product. These are silica gel, fused quartz, fumed silica, and aerogels. It is used in microelectronics, structural materials, and as components in the food and pharmaceutical industries [39].

LaB$_6$-SiO$_2$ composites have excellent NIR absorption and photothermal conversion property [40]. Lai *et al.* synthesized LaB$_6$@-iO$_2$/Fe$_3$O$_4$ composite nanoparticles as a novel nanomaterial for the NIR photothermal ablation of bacteria. From TEM, XRD pattern, and the analyses of absorption spectra, the formation of LaB$_6$-SiO$_2$/Fe$_3$O$_4$ composite nanoparticles was verified by Lai. Van. The LaB$_6$-SiO$_2$/Fe$_3$O$_4$ composite nanoparticles possessed superparamagnetic features and conserved the perfect NIR photothermal conversion feature of LaB$_6$ nanoparticles. The binding of Fe$_3$O$_4$ nanoparticles on the surface of Van-LaB$_6$-SiO$_2$ composite nanoparticles can be seen in Fig. (6.7). TEM images exhibit that particle accumulation could consist of the LaB$_6$-SiO$_2$/Fe$_3$O$_4$ composite nanoparticles [41].

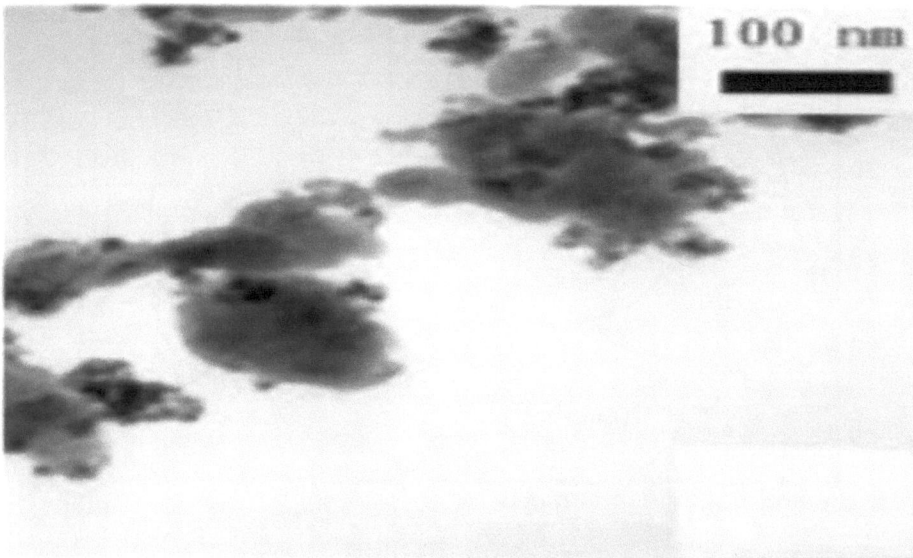

Fig. (6.7). TEM images of Van- LaB$_6$@SiO$_2$/Fe$_3$O$_4$ composite nanoparticles [41] (Copyright © 2013 Acta Materialia Inc.).

Tang *et al.* synthesized LaB_6-SiO_2 nanoparticles (NPs) with a core-shell structure. The thermogravimetric analysis (TGA) and UV-vis-NIR results indicate that LaB_6-SiO_2 NPs contributed to developing thermal stability of PVB in LaB_6-SiO_2-PVB film, with an enhancement of approximately 60 °C from the first decomposition temperature in the state of the clear PVB film. In addition, maximum absorption in the near-infrared region acted lightly from 1466 to 1472 nm and maximum transmittance in the visible region red-shifted from 605 to 645 nm. The LaB_6-SiO_2 NPs can be a potential candidate for heat insulation and transparent materials.

The transmission spectra of 0.30 wt.% LaB_6-SiO_2-PVB and 0.30 wt.% LaB_6-PVB nanocomposite films in the region of 380-2000 nm are shown in Fig. **(6.8)**. It is shown that two specimens have transmittance spectrum curves. These specimens possess strong absorption in the near-infrared region of 780-1500 nm and high transmittance in the visible region of 380-780 nm. The LaB_6-SiO_2-PVB nanocomposite films demonstrated an approximately similar transmittance spectrum with maximum absorption at about 1000 and 1500 nm as the LaB_6 exhibited [42].

Fig. (6.8). NIR-vis- UV transmittance spectra of LaB_6@SiO_2-PVB and LaB_6-PVB nanocomposite films [42] (Copyright© 2014 Elsevier Ltd.).

6.6. CEB$_6$-AL COMPOSITES

Aluminum has a density lower than those of other general metals, at about one-third that of steel. It has a big affinity toward oxygen and creates a protective layer of oxide on the surface when exposed to air. Aluminum visually resembles silver, both in its great ability to reflect light and in its color. It is ductile, non-magnetic, and soft. Aluminum is the most abundant metallic element in Earth's crust [43]. CeB$_6$ and Al have different particle sizes. CeB$_6$ has good chemical stability, a high melting point, and high hardness. That's why CeB$_6$ might be a good boride-type inoculant for aluminum alloys.

The CeB$_6$/Al composite is very smooth and clean [44]. Zhao *et al.* fabricated CeB$_6$/Al composite by induction melting. CeB$_6$ was produced in Al melt at a low temperature. CeB$_6$/Al composite could be refined to nano-scale by melt spinning and exhibit controllable properties. The grain refining impact of CeB$_6$/Al inoculant is openly improved by microstructure control. An inoculation model based on nano-particle clusters was recommended by Zhao. The sub-micron CeB$_6$ clusters are de-agglomerated to lesser ones consisting of nanoparticles with sizes of ~10 nm. The nano-CeB$_6$ particles are immensely in dispersion, resulting in a powerful reinforcement impact.

The TEM image of the Al alloy after inoculation by alloy C3. The size of the dispersion particles is approximately 5 nm which is lesser than that in inoculant ribbons (~100 nm), as shown in Fig. (6.9) [45].

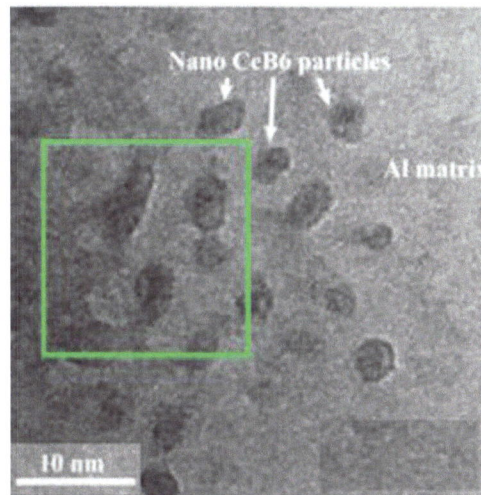

Fig. (6.9). TEM micrograph exhibiting nano-sized CeB$_6$ particles cluster and TEM characterization of the as-cast aluminum inculcated by CeB$_6$/Al ribbon C3 [45] (Copyright © 2017 Elsevier B.V.).

In a different study, Kewu *et al.* produced B_4C-CeB_6/Al composites by pressureless infiltration technology. They investigated the mechanical properties of B_4C-CeB_6/Al composites. The flexibility strength, density, and fracture toughness of the composites were immensely enhanced compared with B_4C, but the hardness of the composites was decreased. The fracture toughness and the flexibility strength of the composites were immensely enhanced for two major reasons. Firstly the fracture toughness and the flexibility strength of the composites performed were enhanced for CeB_6. Secondly, the ductility of aluminum appeared in B_4C-CeB_6/Al composites [46].

6.7. REB$_6$-CARBON NANOTUBES

Carbon nanotubes (CNT) are carbonated structures with a length of up to 1 mm and an external diameter of 4~30 nm and are shaped like honeycombs [47]. CNTs are unique nano-structures with very important mechanical and electronic properties. In addition to nano-sized electronic applications, its various applications have increased interest in their potential [48]. REB$_6$-CNT composites have high emission current density and high brightness, leading to a good coherence electron beam [49].

6.7.1. LAB$_6$-CNT

LaB$_6$'s low operating function can be adapted to the high size ratio of CNTs and is then expected to contribute to improved field emissions [50]. Wei *et al.* replaced the single multi-walled CNT emitter tip with LaB$_6$ rather than CNT-emitter arrays. They reported reduced electric field and decreasing current density [49]. In the study conducted by Kumari *et al.*, the field emissions of CNTs were significantly improved with LaB$_6$ nanoparticles (NPs). CNTs were enlarged by chemical vapor accumulation on the silicon substrate. Physical, morphological, elementary, and graphical nature changes were determined by SEM, TEM, EDS, and Raman analysis. They showed that LaB$_6$-NPs reduced the electrical field from 3.0 to 2.1 V / lm in decorated CNTs [51].

Detailed results from field emission studies of LaB$_6$-coated multi-walled carbon nanotube (MWCNT) films, intact LaB$_6$ films and intact MWCNT films were reported by Patra *et al.* They synthesized the films with a combination of chemical and physical deposition processes. In LaB$_6$ coated MWCNTs, an impressive increase in area enhancement factor and temporal stability compared to intact MWCNT (multi-walled carbon nanotube) and intact LaB$_6$ films, as well as a reduction in opening area and threshold area Surface morphology of the films was examined by SEM. The presence of LaB$_6$ nanoparticles on the outer walls of the CNT's LaB_6-coated MWCNT films was confirmed by SEM. The presence of LaB_6 was confirmed by XPS results and also confirmed by the Raman spectrum. Surface morphologies of unspoiled MWCNT and LaB_6 coated MWCNT composite films were analyzed from FESEM images (Fig. **6.10**). The growth of LaB_6 nanoparticles at the top and walls of MWCNT causes an increase in the effective oscillation area, which is shown in Fig. (**6.10a**). The cultivated CNTs are seen in Fig. (**6.10b**) [52].

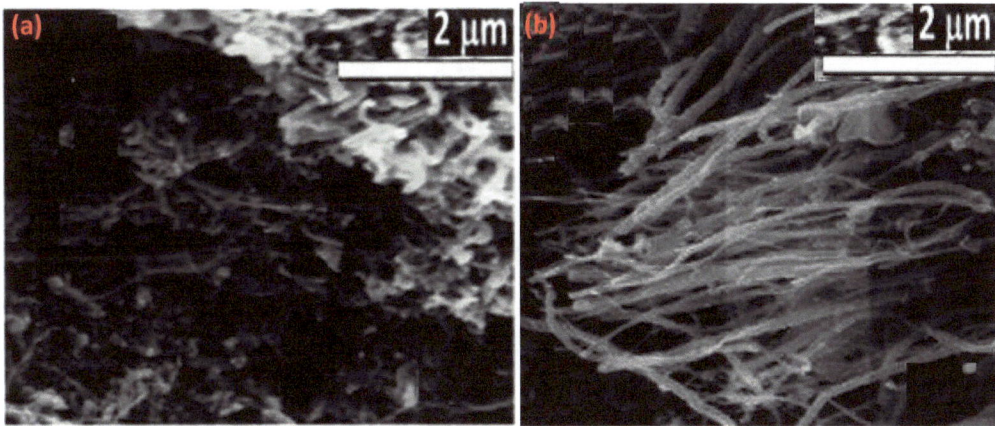

Fig. (6.10). (a) Adopted from the LaB_6-coated CNT film and **(b)** SEM micrograph of an intact MWCNT film [52] (Copyright © 2014 AIP Publishing LLC).

6.7.2. CEB_6-CNT

CeB_6-CNT has a good maximum current density, field electron emission, and field enhancement factor [53]. A detailed FE analysis on CeB_6-coated CNT films grown by combining improved plasma vapor deposition (MPE CVD) and the hydrothermal synthesis routes with microwave plasma was carried out by Jha *et al.* It was observed that a significant increase in the threshold area [54]. In a different study, Patra *et al.* reported the condensed field emission from CeB_6-coated CNTs on the Si substrate and that the new composite material should be proposed as a potential candidate for next-generation electron sources. The film, with a combination of chemical and physical accumulation processes, was synthesized. The CeB_6-coated CNT film showed a significant increase in maximum current density and field development factor compared to pure current CeB_6.

In addition, they observed a decrease in the opening area and threshold area. The elemental composition and surface morphology of the films was seen with SEM, TEM, and EDS measurements. The CeB_6 nanoparticles on the CNTS walls were seen in the TEM images (Fig. **6.11a**) and the FESEM (Fig. **6.11b**) reveal the growth of CeB_6 nanoparticles at the ends and sides of the CNT walls [53].

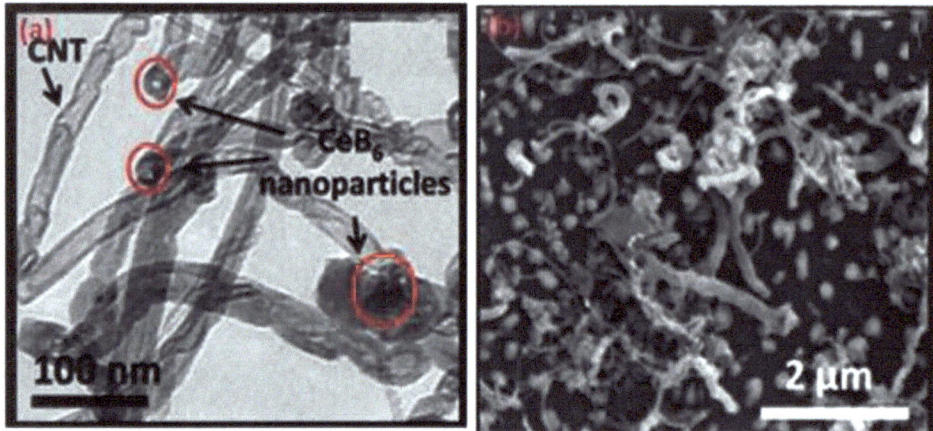

Fig. (6.11). (a) TEM image of CeB_6 coated CNT film with arrows marking CeB_6 nanoparticles in CNT walls **(b)** FESEM images of CeB_6 coated CNT film. Adopted from [53] (Copyright © 2019 AIP Publishing LLC).

6.8. LAB_6-ALUMINA (AL_2O_3) COMPOSITES

Alumina is a widely used material due to its high abrasion resistance, good corrosion resistance at high temperatures, high melting temperature, low density, good mechanical properties, and low cost. However, there are disadvantages. Alumina ceramic has a brittle material structure due to its low fracture toughness. Another disadvantage of alumina is that the application temperatures are limited. Many researchers aimed to create a new composite material by reinforcing it with a different material to eliminate the undesirable properties of alumina [55, 56].

LaB_6-Al_2O_3 ceramic powders may be suitable candidates for particle reinforcement and may contribute to the microstructural and mechanical properties of ceramic or metallic matrix composites. In this study by Tekoglu *et al.*, the pure LaB_6-Al_2O_3 nano-composite powders were synthesized for the first time using a method consisting of high-energy ball milling, annealing, and leaching.

LaB_6-Al_2O_3 ceramic powders may be suitable candidates for particulate supplementation and may contribute to the microstructural and mechanical properties of ceramic or metallic matrix composites. In this study conducted by Tekoğlu and others, they synthesized pure LaB_6-Al_2O_3 nano composite powders for the first time with a method consisting of high-energy ball grinding, pick-up, and lychee processes. They examined the effect of milling time on the formation of LaB_6 and Al_2O_3 phases in terms of detailed microstructural features [57].

6.9. LAB$_6$-PVB NANOCOMPOSITE

Polyvinyl Butyral (PVB) is a kind of polymer that could withstand oxidization impact, wear, and heat. That's why laminated glass is widely used as a thin film. A lot of work has been done to search for composites made up of PVB and nanomaterials [58]. LaB_6-PVB/nanocomposite films could successfully absorb NIR and UV from sunlight, reducing the temperature in the film-covered containers. In addition, they enhance tensile strength.

Zhang *et al.* have successfully produced LaB_6-PVB/nanocomposite films through the technique of solution blending. FESEM was used to observe the distribution status of LaB_6 nanoparticles at the PVB matrix. The NIR -VIS- UV permeability spectrogram of films has shown the effect of LaB_6 on PVB optical properties. The bending test of composite films proved that the strength of composite films had increased somewhat, as LaB_6 adds. The UV-VIS-NIR permeability spectrogram of the LaB_6-contribution PVB is based on pure PVB film. The five-additive spectrum had a similar shape, and as LaB_6 content increased, light transmission decreased through composite films. It was remarkable that important peaks around 225 nm, 1700 nm, and 2300 nm wavelengths appeared on all curves except the pure PVB curve, which meant that composite films had a good ability to absorb UV and NIR. LaB_6 nanoparticles can slightly affect VIS's transmission, and light with a wavelength of about 600 nm has the biggest permeability to show composite films have the slightest impact on yellow light [59]. In another study, Tang *et al.* successfully fabricated LaB_6-PVB composites by a solution casting method. TGA and UV-vis-NIR results reveal that nanocomposite films have exceptional thermal stability. The temperature where 5 % weight loss of the PVB matrix was developed after the addition of LaB_6 and Tin-doped Indium oxide (ITO) nanoparticles and the property for blocking near-infrared light was also improved as compared with the state of pure PVB film [60].

6.10. LAB$_6$-PMMA COMPOSITE

Polymethyl methacrylate (PMMA) is a widely used low-cost thermoplastic polymer with unlimited applications for daily life. High transparency makes PMMA a good replacement for glass where weight or impact is an important concern. PMMA is harmonious with human tissue making it a significant material for prosthetics and transplants, particularly in the field of ophthalmology due to its transparent properties [61].

The LaB_6- PMMA composite has a good optical property, enhancing light absorption strength. In addition, the composite has the best performance in NIR and VIS absorption. Yuan *et al.* successfully produced the modified LaB_6 particles in sizes ranging from 50 nm to 400 nm were added to the PMMA matrix to look for the effect of LaB_6 particles added on the optical properties of LaB_6/PMMA. Ultraviolet-visible-near-infrared (UV-vis-NIR) absorption spectrum was used to examine the optical properties of the as-prepared materials. The difference in particle size can affect the absorption of composites around 600 nm wavelength of visible light. PMMA doped with LaB_6 had more powerful light absorption than pure PMMA [62]. In another study, Tang *et al.* successfully synthesized the LaB_6-PMMA composite by a solid-state reaction in a vacuum. A silane coupling agent is used to change the surface of the nanoparticles. This process is carried out in dissimilar solutions to determine the best modification technology. The modified LaB_6 nanoparticles are added to the PMMA matrix *via in-situ* polymerization. The synthesized nanoparticles are defined by TEM and XRD, the results indicate that the resulting particles have a high degree of elliptical and crystallinity or cubic shape, with a size of between 20 nm and 100 nm [63]

6.11. LAB_6-$MOSI_2$-SIC COMPOSITES

Molybdenum disilicide (molybdenum silicide or $MoSi_2$) is a refractory ceramic with primary use in heating elements. $MoSi_2$ is electrically conductive and has moderate density, and its melting point is 2030 °C. $MoSi_2$ is a gray metallic-looking material with a tetragonal crystal structure (alpha-modification), and also its beta-modification is hexagonal and unstable. $MoSi_2$ is soluble in hydrofluoric acid and nitric acid but insoluble in most acids [64].

The LaB_6-$MoSi_2$-SiC composite coating has an intense structure and excellent thermal shock resistance. In addition, the composite increases fracture toughness, compactness, and flexural strength. Li *et al.* successfully fabricated LaB_6-$MoSi_2$-SiC and $MoSi_2$-SiC coatings on the surface of carbon/carbon composites by the pack cementation method. The crystal structures of the LaB_6-$MoSi_2$-SiC coatings were investigated by XRD. The element distributions and morphologies were also examined by energy dispersive spectroscopy (EDXS) and SEM. The impact of LaB_6 on thermal shock resistance and the microstructure of $MoSi_2$-SiC coating was examined. The results showed that the LaB_6-$MoSi_2$-SiC composites possessed superior thermal shock resistance and a denser structure. The weight losses of LaB_6-$MoSi_2$-SiC (LMS) and $MoSi_2$-SiC (MS) coated samples have been between 0.627% and 2.019%, respectively, after 25 times of thermal cycle oxidation between 1773 K and room temperature. In this study, the weight loss rate of the LaB_6-$MoSi_2$-SiC coated sample is lower than that of the $MoSi_2$-SiC coated specimen. The weight loss rate of the LaB_6-$MoSi_2$-SiC coated specimen is 1,1 g/m^2 time. Furthermore, the weight loss rate of the $MoSi_2$-SiC coated specimen increases to 2,1 g/m^2 time. That indicates that the LaB_6-$MoSi_2$-SiCcomposites specimen shows a better thermal stress resistance [65]. Wang *et al.* have successfully synthesized LaB_6-$MoSi_2$-SiC composited by the combination of multiple methods, including pack cementation, supersonic atmospheric plasma spraying, and CVD. Isothermal oxidation results showed that after oxidation for 200 h at 1773 K, the mass loss of LaB_6-$MoSi_2$-SiC composites coating decreased from $4.34 \pm 0.28\%$ to $1.12 \pm 0.23\%$. The isothermal oxidation behavior of LMS is shown in Fig. (**6.12**). Before the involvement of SiC-nanowires in the LMS/SiC composites, no significant mass acquisition process had occurred, and the overall oxidation process of the LMS/SiC composites sample indicated a mass loss behavior with an increasing parabolic model, and the final mass loss reached $4.34 \pm 0.28\%$ after 200 hours of oxidation at 1773 K [66].

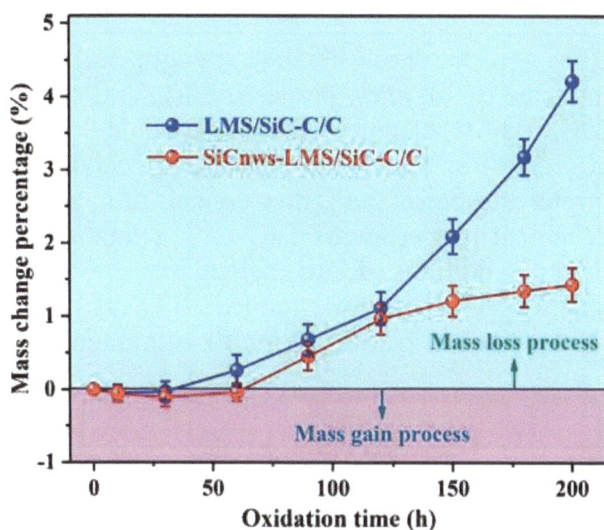

Fig. (6.12). Isothermal oxidation curves of the samples with LaB_6-$MoSi_2$-SiC composites at 1773 K in static. Adopted from [66] (Copyright© 2018 Elsevier Ltd and Techna Group S.r.l.).

6.12. REB_6-OTHER COMPOSITES

LaB_6@SiO_2/Au composite has superb NIR absorption and photothermal conversion features. This composite induces heating of the reaction medium and also provides hot spots on the catalyst surface. Lai *et al.* successfully synthesized LaB_6@SiO_2/Au composite nanoparticles. The TEM images of LaB_6 and LaB_6@SiO_2/Au nanoparticles are shown in Fig. (**6.13**). The silica was coated on the surface of LaB_6 nanoparticles [40].

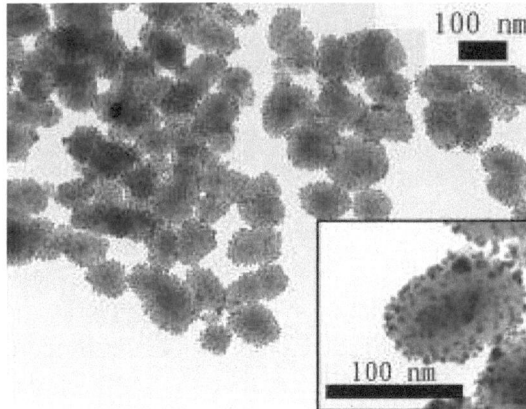

Fig. (6.13). TEM images of and LaB_6@SiO_2/Au composite nanoparticles. Adopted from [40] (Copyright © 2013 Elsevier B.V.).

LaB_6-TaC composite has good hardness, bending strength, and fracture toughness. A different study conducted by Lu *et al.* prepared LaB_6-TaC composite ceramics with hot pressing. LaB_6 may react with Tantalum carbide TaC to facilitate the intensification of TaC ceramics. They examined the effects of LaB_6 addition on phase composition, microstructure, and mechanical properties of TaC-LaB_6 composites. The mechanical properties of TaC-LaB_6 composites have changed according to LaB_6 content. Additions of LaB_6 helped improve the densification of the TaC-based composites [67]. The formation of B_2O_3 as a minor phase was shown by TEM. Many B_2O_3 were located on multigrain junctions and are shown in Fig. (**6.14**).

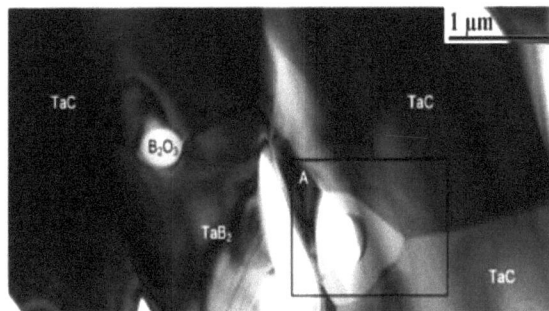

Fig. (6.14). TEM micrograph of the TaC ceramic with 1 mol% LaB_6, indicating B_2O_3 at both multigrain junction boundary and grain boundaries. Adopted from [67] (Copyright© 2018 Elsevier B.V.).

CONCLUSION

The REB_6 composites can also be used as a good structural material at high temperatures owing to their high melting points, excellent strength, and creep resistance. REB_6 is composited with some kind of ceramics, leading to significant properties, such as strong hardness and friction resistance at high temperatures. The sizes of these materials are also significant as nano, and micro-sized ceramic materials have different types of properties when forming a composite with REB_6 or any materials.

REFERENCES

[1]   Bogomol, I.; Nishimura, T.; Nesterenko, Y.; Vasylkiv, O.; Sakka, Y.; Loboda, P. The bending strength temperature dependence of the directionally solidified eutectic LaB_6–ZrB_2 composite. *J. Alloys Compd.,* **2011**, *509*(20), 6123-6129.
[http://dx.doi.org/10.1016/j.jallcom.2011.02.176]

[2]   Nesmelov, D.D.; Vikhman, S.V.; Novoselov, E.S.; Perevislov, S.N.; Ordan'yan, S.S. Structure, hardness and fracture toughness of arc-melted LaB_6-TiB_2 eutectic alloy. *IOP Conf. Series Mater. Sci. Eng.,* **2019**, *525*(1), 012066.
[http://dx.doi.org/10.1088/1757-899X/525/1/012066]

[3]   Taran, A.; Voronovich, D.; Oranskaya, D.; Filipov, V.; Podshyvalova, O. Thermionic emission of LaB_6–ZrB_2 quasi-binary eutectic alloy with different ZrB_2 fibers orientation *Funct. Mater.,* **2013**.
[http://dx.doi.org/10.15407/fm20.04.485]

[4]   Paderno, Y.B.; Paderno, V.N.; Filippov, V.B.; Mil'man, Y.V.; Martynenko, A.N. "Structure features of the eutectic alloys of borides with the d-and f-transition metals," *Sov. powder Metall. Met. Ceram.,* **1992**, *31*(8), 700-706.

[5]   Yang, X.; Wang, P.; Wang, Z.; Hu, K.; Cheng, H.; Li, Z.; Zhang, J. Microstructure, mechanical and thermionic emission properties of a directionally solidified LaB_6-VB_2 eutectic composite. *Mater. Des.,* **2017**, *133*, 299-306.
[http://dx.doi.org/10.1016/j.matdes.2017.07.069]

[6]   Taran, A.; Voronovich, D.; Plankovskyy, S.; Paderno, V.; Filipov, V. Review of LaB_6S, Re-W Dispenser, and $BaHfO_3S$-W Cathode Development. *IEEE Trans. Electron Dev.,* **2009**, *56*(5), 812-817.
[http://dx.doi.org/10.1109/TED.2009.2015615]

[7]   Ashbrook, R.L. Directionally solidified ceramic eutectics. *J. Am. Ceram. Soc.,* **1977**, *60*(9-10), 428-435.
[http://dx.doi.org/10.1111/j.1151-2916.1977.tb15527.x]

[8]   Waku, Y.; Nakagawa, N.; Wakamoto, T.; Ohtsubo, H.; Shimizu, K.; Kohtoku, Y. A ductile ceramic eutectic composite with high strength at 1,873 K. *Nature,* **1997**, *389*(6646), 49-52.
[http://dx.doi.org/10.1038/37937]

[9]   Dickey, E.C.; Dravid, V.P.; Nellist, P.D.; Wallis, D.J.; Pennycook, S.J. Three-dimensional atomic structure of NiO–ZrO_2(cubic) interfaces. *Acta Mater.,* **1998**, *46*(5), 1801-1816.
[http://dx.doi.org/10.1016/S1359-6454(97)00373-X]

[10]   Loboda, P.I. Features of structure formation with zone melting of powder boron-containing refractory materials. *Powder Metall. Met. Ceramics,* **2000**, *39*(9/10), 480-486.
[http://dx.doi.org/10.1023/A:1011322707881]

[11]   Paderno, Y.; Paderno, V.; Filippov, V. Some peculiarities of eutectic crystallization of LaB_6–(Ti, Zr) B_2 alloys. *J. Solid State Chem.,* **2000**, *154*(1), 165-167.
[http://dx.doi.org/10.1006/jssc.2000.8830]

[12]   Chen, C.M.; Zhang, L.T.; Zhou, W.C. Characterization of LaB_6–ZrB_2 eutectic composite grown by the floating zone method. *J. Cryst. Growth,* **1998**, *191*(4), 873-878.
[http://dx.doi.org/10.1016/S0022-0248(98)00358-3]

[13]   Schwetz, K. A. Boron carbide, boron nitride, and metal borides. *Ullmann's Encycl. Ind. Chem.,* **1985**.

[14]   Kuznetsov, G. High temperature cathodes for high current density. *Nucl. Instrum. Methods Phys. Res. A,* **1994**, *340*(1), 204-208.
       [http://dx.doi.org/10.1016/0168-9002(94)91302-1]

[15]   Wang, S.C.; Wei, W.C.J.; Zhang, L.T. Microstructural characterization of LaB_6-ZrB_2 eutectic composites. *Key Eng. Mater.,* **2003**, *249*, 101-104.
       [http://dx.doi.org/10.4028/www.scientific.net/KEM.249.101]

[16]   Min, G.H.; Gao, R.; Yu, H.S.; Han, J. Mechanical properties of LaB_6-ZrB_2 composites. *Key Eng. Mater.,* **2005**, *297-300*, 1630-1638.
       [http://dx.doi.org/10.4028/www.scientific.net/KEM.297-300.1630]

[17]   Gao, R.; Min, G.; Yu, H.; Zheng, S.; Lu, Q.; Han, J.; Wang, W. Fabrication and oxidation behavior of LaB_6–ZrB_2 composites. *Ceram. Int.,* **2005**, *31*(1), 15-19.
       [http://dx.doi.org/10.1016/j.ceramint.2004.02.006]

[18]   Sokol, I.V.; Krasnova, T.V. Composition of titanium boride synthesis products. *Ind. Lab. Transl. Transl.,* **1993**, *59*(6), 564-567.

[19]   Martienssen, W.; Warlimont, H. *Springer Handbook of Condensed Matter and Materials Data*; Springer Science & Business Media, **2006**.

[20]   Bogomol, I.; Nishimura, T.; Vasylkiv, O.; Sakka, Y.; Loboda, P. High-temperature strength of directionally reinforced LaB_6–TiB_2 composite. *J. Alloys Compd.,* **2010**, *505*(1), 130-134.
       [http://dx.doi.org/10.1016/j.jallcom.2010.05.003]

[21]   Soloviova, T.O.; Karasevska, O.P.; Vleugels, J.; Loboda, P.I. Influence of annealing on crucible-free float zone melted LaB_6-TiB_2 composites. *J. Alloys Compd.,* **2017**, *729*, 749-757.
       [http://dx.doi.org/10.1016/j.jallcom.2017.09.223]

[22]   Murarka, S.P. Refractory silicides for integrated circuits. *J. Vac. Sci. Technol.,* **1980**, *17*(4), 775-792.
       [http://dx.doi.org/10.1116/1.570560]

[23]   Sonber, J.K.; Murthy, T.S.R.C.; Sairam, K.; Paul, B.; Chakravartty, J.K. Effect of TiSi2 addition on densification of Cerium hexaboride. *Ceram. Int.,* **2016**, *42*(1), 891-896.
       [http://dx.doi.org/10.1016/j.ceramint.2015.09.015]

[24]   Yu, J.; Ma, L.; Abbas, A.; Zhang, Y.; Gong, H.; Wang, X.; Zhou, L.; Liu, H. Carbothermal reduction synthesis of TiB_2 ultrafine powders. *Ceram. Int.,* **2016**, *42*(3), 3916-3920.
       [http://dx.doi.org/10.1016/j.ceramint.2015.11.059]

[25]   Xiao, L.; Su, Y.; Ran, J.; Liu, Y.; Qiu, W.; Wu, J.; Lu, F.; Shao, F.; Tang, D.; Peng, P. First-principles prediction of solar radiation shielding performance for transparent windows of GdB_6. *J. Appl. Phys.,* **2016**, *119*(16), 164903.
       [http://dx.doi.org/10.1063/1.4945679]

[26]   Ordan'yan, S.S.; Nikolaeva, E.E. Interaction in the GdB_6-TiB_2 system. *Soviet Powder Metallurgy and Metal Ceramics,* **1987**, *26*(1), 51-53.
       [http://dx.doi.org/10.1007/BF00794265]

[27]   Zoli, L.; Galizia, P.; Silvestroni, L.; Sciti, D. Synthesis of group IV and V metal diboride nanocrystals *via* borothermal reduction with sodium borohydride. *J. Am. Ceram. Soc.,* **2018**, *101*(6), 2627-2637.
       [http://dx.doi.org/10.1111/jace.15401]

[28]   Ordan'yan, S.S.; Nikolaeva, E.E.; Martynova, T.Y. Interaction in the system GdB_6-HfB_2. *Zhurnal Neorg. Khimii,* **1987**, *32*(11), 2864-2866.

[29]   Pierson, H.O. *Handbook of Refractory Carbides & Nitrides: Properties, Characteristics, Processing and Apps*; William Andrew, **1996**.

[30]   Wang, F.; Cheng, L.; Xie, Y.; Jian, J.; Zhang, L. Effects of SiC shape and oxidation on the infrared

emissivity properties of ZrB_2–SiC ceramics. *J. Alloys Compd.,* **2015**, *625*, 1-7.
[http://dx.doi.org/10.1016/j.jallcom.2014.09.191]

[31]   Asl, M.S.; Kakroudi, M.G.; Noori, S. Hardness and toughness of hot pressed ZrB_2–SiC composites consolidated under relatively low pressure. *J. Alloys Compd.,* **2015**, *619*, 481-487.
[http://dx.doi.org/10.1016/j.jallcom.2014.09.006]

[32]   Akin, I.; Hotta, M.; Sahin, F.C.; Yucel, O.; Goller, G.; Goto, T. Microstructure and densification of ZrB_2–SiC composites prepared by spark plasma sintering. *J. Eur. Ceram. Soc.,* **2009**, *29*(11), 2379-2385.
[http://dx.doi.org/10.1016/j.jeurceramsoc.2009.01.011]

[33]   Tian, W.B.; Kan, Y.M.; Zhang, G.J.; Wang, P.L. Effect of carbon nanotubes on the properties of ZrB_2–SiC ceramics. *Mater. Sci. Eng. A,* **2008**, *487*(1-2), 568-573.
[http://dx.doi.org/10.1016/j.msea.2007.11.027]

[34]   Yang, X.; Wang, X.; Wang, P.; Hu, K.; Li, Z.; Zhang, J. Spark plasma sintering of SiC-LaB_6 composite. *J. Alloys Compd.,* **2017**, *704*, 329-335.
[http://dx.doi.org/10.1016/j.jallcom.2017.02.033]

[35]   Zhu, Q.; Oganov, A.R.; Lyakhov, A.O. Novel stable compounds in the Mg–O system under high pressure. *Phys. Chem. Chem. Phys.,* **2013**, *15*(20), 7696-7700.
[http://dx.doi.org/10.1039/c3cp50678a] [PMID: 23595296]

[36]   Shand, M.A. *The chemistry and technology of magnesia*; Vol. *210*, **2006**.
[http://dx.doi.org/10.1002/0471980579]

[37]   Deng, J.; Zeng, B.Q.; Wang, X.J.; Lin, Z.L.; Qi, K.C.; Cao, G.C. Improvement of Discharge Properties by Printing \$LaB_6\$ Films on MgO Protective Layers in Plasma Display Panel. *IEEE Electron Device Lett.,* **2013**, *34*(8), 1026-1028.
[http://dx.doi.org/10.1109/LED.2013.2265412]

[38]   Deng, J.; Zeng, B.; Wang, Y.; Wang, X.; Lin, Z.; Qi, K.; Cao, G. Fabrication and characteristics of LaB_6-doped MgO protective layer for plasma display panels. *Mater. Lett.,* **2014**, *134*, 51-55.
[http://dx.doi.org/10.1016/j.matlet.2014.07.062]

[39]   Iler, K. R. The Chemistry of Silica. In: *Solubility, Polym. Colloid Surf. Prop. Biochem. Silica,* **1979**.

[40]   Lai, B.H.; Lin, Y.R.; Chen, D.H. Fabrication of LaB_6@SiO_2/Au composite nanoparticles as a catalyst with near infrared photothermally enhanced activity. *Chem. Eng. J.,* **2013**, *223*, 418-424.
[http://dx.doi.org/10.1016/j.cej.2013.02.109]

[41]   Lai, B.H.; Chen, D.H. Vancomycin-modified LaB_6@SiO_2/Fe_3O_4 composite nanoparticles for near-infrared photothermal ablation of bacteria. *Acta Biomater.,* **2013**, *9*(7), 7573-7579.
[http://dx.doi.org/10.1016/j.actbio.2013.03.023] [PMID: 23535232]

[42]   Tang, H.; Su, Y.; Tan, J.; Hu, T.; Gong, J.; Xiao, L. Optical properties and thermal stability of Poly(vinyl butyral) films embedded with LaB_6@SiO_2 core–shell nanoparticles. *Superlattices Microstruct.,* **2014**, *75*, 908-915.
[http://dx.doi.org/10.1016/j.spmi.2014.09.020]

[43]   Lide, D.R. Magnetic susceptibility of the elements and inorganic compounds. *Handb. Chem. Phys.,* **2005**, *81*, 130-135.

[44]   Liu, S.; Wang, X.; Cui, C.; Zhao, L.; Liu, S.; Chen, C. Fabrication, microstructure and refining mechanism of *in situ* CeB_6/Al inoculant in aluminum. *Mater. Des.,* **2015**, *65*, 432-437.
[http://dx.doi.org/10.1016/j.matdes.2014.09.038]

[45]   Liu, S.; Wang, X.; Cui, C.; Zhao, L.C.; Li, N.; Zhang, Z.; Ding, J.; Sha, D. Enhanced grain refinement of *in situ* CeB_6/Al composite inoculant on pure aluminum by microstructure control. *J. Alloys Compd.,* **2017**, *701*, 926-934.
[http://dx.doi.org/10.1016/j.jallcom.2017.01.188]

[46] Kewu, P.; Wenyuan, W.; Jingyu, X.; Ganfeng, T.; Fuhu, N. Study on Mechanical Properties and Fracture Mechanisms of B4 C-CeB$_6$/A1 Composites. *J. Rare-Earths,* **2007**, *25*, 77-81.
[http://dx.doi.org/10.1016/S1002-0721(07)60528-6]

[47] De Volder, M.F.L.; Tawfick, S.H.; Baughman, R.H.; Hart, A.J. Carbon nanotubes: present and future commercial applications, **2013**.

[48] Avouris, P.; Dresselhaus, G.; Dresselhaus, M.S. Topics in Applied Physics. In: *Carbon nanotubes: synthesis, structure, properties and applications*, **2000**.

[49] Wei, W.; Jiang, K.; Wei, Y.; Liu, P.; Liu, K.; Zhang, L.; Li, Q.; Fan, S. LaB$_6$ tip-modified multiwalled carbon nanotube as high quality field emission electron source. *Appl. Phys. Lett.,* **2006**, *89*(20), 203112.
[http://dx.doi.org/10.1063/1.2388862]

[50] Wang, X.; Lin, Z.; Qi, K.; Chen, Z.; Wang, Z.; Jiang, Y. Field emission characteristics of lanthanum hexaboride coated silicon field emitters. *J. Phys. D Appl. Phys.,* **2007**, *40*(16), 4775-4778.
[http://dx.doi.org/10.1088/0022-3727/40/16/006]

[51] Kumari, M.; Gautam, S.; Shah, P.V.; Pal, S.; Ojha, U.S.; Kumar, A.; Naik, A.A.; Rawat, J.S.; Chaudhury, P.K.; Harsh, ; Tandon, R.P. Improving the field emission of carbon nanotubes by lanthanum-hexaboride nano-particles decoration. *Appl. Phys. Lett.,* **2012**, *101*(12), 123116.
[http://dx.doi.org/10.1063/1.4754110]

[52] Patra, R.; Ghosh, S.; Sheremet, E.; Jha, M.; Rodriguez, R.D.; Lehmann, D.; Ganguli, A.K.; Schmidt, H.; Schulze, S.; Hietschold, M.; Zahn, D.R.T.; Schmidt, O.G. Enhanced field emission from lanthanum hexaboride coated multiwalled carbon nanotubes: Correlation with physical properties. *J. Appl. Phys.,* **2014**, *116*(16), 164309.
[http://dx.doi.org/10.1063/1.4898352]

[53] Patra, R.; Ghosh, S.; Sheremet, E.; Jha, M.; Rodriguez, R.D.; Lehmann, D.; Ganguli, A.K.; Gordan, O.D.; Schmidt, H.; Schulze, S.; Zahn, D.R.T.; Schmidt, O.G. Enhanced field emission from cerium hexaboride coated multiwalled carbon nanotube composite films: A potential material for next generation electron sources. *J. Appl. Phys.,* **2014**, *115*(9), 094302.
[http://dx.doi.org/10.1063/1.4866990]

[54] Jha, M.; Patra, R.; Ghosh, S.; Ganguli, A.K. Vertically aligned cerium hexaboride nanorods with enhanced field emission properties. *J. Mater. Chem.,* **2012**, *22*(13), 6356-6366.
[http://dx.doi.org/10.1039/c2jm16538d]

[55] Torralba, J.M.; da Costa, C.E.; Velasco, F. P/M aluminum matrix composites: an overview. *J. Mater. Process. Technol.,* **2003**, *133*(1-2), 203-206.
[http://dx.doi.org/10.1016/S0924-0136(02)00234-0]

[56] Çerezci, T. *"Nikel Partikül Takviyeli Alümina Seramik Kompozitlerin Sentezi ve Karakterizasyonu." Yüksek Lisans Tezi, Sakarya Üniversitesi, Fen Bilimleri Enstitüsü*; Sakarya, **2008**.

[57] Tekoğlu, E.; İmer, C.; Ağaoğulları, D.; Lütfi Öveçoğlu, M. Synthesis of LaB$_6$–Al$_2$O$_3$ nanocomposite powders *via* ball milling-assisted annealing. *J. Mater. Sci.,* **2018**, *53*(19), 13538-13549.
[http://dx.doi.org/10.1007/s10853-018-2454-6]

[58] Gu, F.; Wang, L.; Zhao, Z.; Liu, Y.; Sun, Y.; Yan, L. "Study of poly(vinyl butyral)/silica nanocomposite materials.," *Fuhe Cailiao Xuebao(Acta Mater. Compos. Sin.,* **2003**, *20*, 77-81.

[59] Zhang, L.; Yuan, Y.; Min, G. Preparation and properties of PVB/nano-LaB$_6$ composite films *Proceedings of 2011 International Conference on Electronic & Mechanical Engineering and Information Technology,* **2011**, *vol. 6*, pp. 3127-3129.
[http://dx.doi.org/10.1109/EMEIT.2011.6023049]

[60] Tang, H.; Su, Y.; Hu, T.; Liu, S.; Mu, S.; Xiao, L. Synergetic effect of LaB$_6$ and ITO nanoparticles on optical properties and thermal stability of poly(vinylbutyral) nanocomposite films. *Appl. Phys., A Mater. Sci. Process.,* **2014**, *117*(4), 2127-2132.

[http://dx.doi.org/10.1007/s00339-014-8632-8]

[61]   Brown, R.P. Plastics materials: By JA Brydson. Butterworths Scientific, London **1988**.

[62]   Yuan, Y.; Zhang, L.; Hu, L.; Wang, W.; Min, G. Size effect of added LaB_6 particles on optical properties of LaB_6/Polymer composites. *J. Solid State Chem.,* **2011**, *184*(12), 3364-3367.
        [http://dx.doi.org/10.1016/j.jssc.2011.10.036]

[63]   Hu, L.; Zhang, L.; Yuan, Y.; Min, G. Microstructure and Optical Properties of PMMA Matrix Composites Containing LaB_6 Nanoparticles. *Металлофизика и новейшие технологии,* **2013**.

[64]   d'Heurle, F.M.; Petersson, C.S.; Tsai, M.Y. Observations on the hexagonal form of $MoSi_2$ and WSi_2 films produced by ion implantation and on related snowplow effects. *J. Appl. Phys.,* **1980**, *51*(11), 5976-5980.
        [http://dx.doi.org/10.1063/1.327517]

[65]   Li, T.; Li, H.; Shi, X. Effect of LaB_6 on the thermal shock property of $MoSi_2$-SiC coating for carbon/carbon composites. *Appl. Surf. Sci.,* **2013**, *264*, 88-93.
        [http://dx.doi.org/10.1016/j.apsusc.2012.09.124]

[66]   Wang, C.; Li, K.; He, Q.; Su, Y.; Huo, C.; Shi, X. Oxidation resistance and mechanical properties of LaB_6-$MoSi_2$-SiC ceramic coating toughened by SiC nanowires. *Ceram. Int.,* **2018**, *44*(14), 16365-16378.
        [http://dx.doi.org/10.1016/j.ceramint.2018.06.045]

[67]   Lu, Z.; Liu, L.; Hou, Z.; Geng, G.; Wang, Y.; Sun, W.; Hai, W.; Chen, Y. Effects of LaB_6 content on microstructure and mechanical properties of TaC-LaB_6 composites. *Mater. Chem. Phys.,* **2018**, *213*, 374-382.
        [http://dx.doi.org/10.1016/j.matchemphys.2018.04.060]

CONCLUSION

Rapid progress in several significant branches of modern industry will be directly affected by the development of new or improved technological instruments and machines, and thus, from this perspective, one of particular attention should be dedicated to the manufacture and extensive application of materials with tailor/superior properties, referring to increase in the life and operational reliability of instruments and machines. Certain metal-like compounds, specifically, hexaborides of rare-earth metals, have been seen as suitable for this purpose. With the aid of some of these materials, modern scientific and engineering problems can be handled, and advances in new technological instruments and machines can be moved to the next levels since rare-earth hexaborides are promising materials due to their unique characteristics, such as high melting point, hardness, chemical stability, low work function, low volatility at high temperatures, superconductivity, magnetic properties, efficiency, thermionic emission, and narrowband semiconductivity.

Metal hexaborides, including rare-earth element metals, are a class of simple cubic structured refractory materials with symmetry. This group of metal hexaborides has low electronic work function, low electrical resistance, and thermal expansion coefficient (at some temperature ranges), combined with high hardness and stiffness, high chemical and thermal stability, and melting points. Due to these properties, they can be used in a wide range of industrial applications, ranging from metallurgical to the electronics industry. High-resolution optical systems, welding technology, detectors, and metallic coatings based on high voltage and temperature, thermionic materials, electron microscopes, X-ray tubes, and nuclear materials are some of these areas. Furthermore, the use of metal hexaborides as a component of composite materials in the aerospace industry has gradually increased. To understand and improve the chemical and physical properties of the material, it is necessary to use the correct methods with high-purity production, and then the electronic properties of these structures should be investigated. Many production techniques have been applied in the preparation of rare-earth hexaborides. Traditionally, rare-earth borides are synthesized by high-temperature reaction processes, such as the direct solid-phase reaction of the corresponding elements/compounds, the carbothermal reaction of the rare-earth oxides, and B or boron carbide (B_4C). As a result, after the successful production of these materials, very fine powders with a high purity level can be produced homogeneously. After a successful production, microstructural, electronic, optical, mechanical, and thermal analyses should be examined. After these results, one of the doping, alloying, and compositing processes will be done to improve the properties of the materials. Before synthesizing rare-earth metal hexaborides, which materials and at what rate should be doped, alloyed, and composited should be reviewed.

This book presents an overview of synthesizes, properties, and application areas of rare-earth metal hexaborides to guide researchers and engineers in pursuing these interesting and unique materials. Also, this study focuses on recent developments and trends regarding the synthesis, characterization, and applications of these materials.

SUBJECT INDEX

www.ingramcontent.com/pod-product-compliance
Lightning Source LLC
Chambersburg PA
CBHW041716210326
41598CB00007B/670